任务型语码转换式**双语**教学系列教材

总主编 刘玉彬 副总主编 杜元虎 总主审 段晓东

机械工程

MECHANICAL ENGINEERING

主 编 康 晶 白 兰
主 审 梁艳君

大连理工大学出版社

图书在版编目(CIP)数据

机械工程 / 康晶,白兰主编. — 大连：大连理工大学出版社,2014.6
任务型语码转换式双语教学系列教材
ISBN 978-7-5611-9136-1

Ⅰ.①机… Ⅱ.①康… ②白… Ⅲ.①机械工程－双语教学－高等学校－教材－英、汉 Ⅳ.①TH

中国版本图书馆 CIP 数据核字(2014)第 095298 号

大连理工大学出版社出版

地址：大连市软件园路80号　邮政编码：116023
发行：0411-84708842　邮购：0411-84703636　传真：0411-84701466
E-mail:dutp@dutp.cn　URL:http://www.dutp.cn
大连理工印刷有限公司印刷　　大连理工大学出版社发行

幅面尺寸：183mm×233mm　　印张：10.75　　字数：350千字
2014年6月第1版　　　　　　2014年6月第1次印刷

责任编辑：邵　婉　　　　　　　责任校对：诗　宇
　　　　　　　　封面设计：波　朗

ISBN 978-7-5611-9136-1　　　　　　　　定　价：25.00元

2014年的初夏,我们为广大师生奉上这套"任务型语码转换式双语教学系列教材"。

"任务型语码转换式双语教学"是双语教学内涵建设的成果,主要由两大模块构成:课上,以不影响学科授课进度为前提,根据学生实际、专业特点、学年变化及社会需求等,适时适量地渗透英语专业语汇、语句、语段或语篇,"润物细无声"般地扩大学生专业语汇量,提高学生专业英语能力;课外,可向学生提供多种选择的"用中学"平台,如英语科技文献翻译、英语实验报告、英语学术论文、英语小论文、英语课程设计报告、模拟国际研讨会、英语辩论、工作室英语讨论会等,使学生的专业英语实践及应用达到一定频度和数量,激活英语与学科知识的相互渗透,培养学生用英语学习、科研、工作的能力及适应教育国际化和经济一体化的能力。

为保证"任务型语码转换式双语教学"有计划、系统、高效、科学地持续运行,减少教学的随意性和盲目性,方便师生的教与学,我们编写了这套"任务型语码转换式双语教学系列教材"。

本套教材的全部内容均采用汉英双语编写。

教材按专业组册,涵盖所有主干专业课和专业基础课,力求较为全面地反映各学科领域的知识体系。

分册教材编写以中文版课程教材为单位,即一门课为分册教材的一章,每章内容以中文版教材章节为序,每门课以一本中文教材为蓝本,兼顾其他同类教材内容,蓝本教材绝大部分是面向21世纪的国家规划教材。

教材的词汇短语部分,注意体现学科发展的新词、新语,同时考虑课程需求及专业特点,在不同程度上灵活渗透了各章节的重要概念、定义,概述了体现章节内容主旨的语句及语段。分册教材还编写了体现各自专业特点的渗透内容,如例题及解题方法,课程的发生、发展及前沿简介,图示,实验原理,合同文本,案例分析,法条,计算机操作错误提示等。

部分教材补充了中文教材未能体现的先进理论、先进工艺、先进材料或先进方法的核心内容,弥补了某些中文教材内容相对滞后的不足;部分教材概述了各自专业常用研究方法、最新研究成果及学术发展的趋势动态;部分教材还选择性地把编者的部分科研成果转化为教材内容,以期启发学生的创新思维,开阔学生的视野,丰富学生的知识结构,从教材角度支持学生参与科研活动。

本套教材大多数分册都编写了对"用中学"任务实施具有指导性的内容,应用性内容的设计及编写比例因专业而异。与专业紧密结合的应用性内容包括英语写作介绍,如英语实验报告写作,英语论文写作,英语论文摘要写作,英语产品、作品或项目的概要介绍写作等。应用性内容的编写旨在降低学生参与各种实践应用活动的难度,提高学生参与"用中学"活动的可实现性,帮助学生提高完成"用中学"任务的质量水平。

考虑学生英语写作和汉译英的方便,多数分册教材都编写了词汇与短语索引。

"任务型语码转换式双语教学系列教材"尚属尝试性首创,是多人辛勤耐心劳作的结果。尽管在编写过程中,我们一边使用一边修改,力求教材的实用性、知识性、先进性融为一体,希望教材能对学生专业语汇积累及专业资料阅读、英语写作、英汉互译能力的提高发挥作用;尽管编者在教材编写的同时也都在实践"任务型语码转换式双语教学",但由于我们缺乏经验,学识水平和占有资料有限,加上为使学生尽早使用教材,编写时间仓促,在教材内容编写、译文处理、分类体系等方面存在缺点、疏忽和失误,恳请各方专家和广大师生对本套教材提出批评和建议,以期再版时更加完善。

在教材的编写过程中,大量中外出版物中的内容给了我们重要启示和权威性的参考帮助,在此,我们谨向有关资料的编著者致以诚挚的谢意!

编　者

2014 年 5 月

前言 FOREWORD

双语教学是我国高等教育教学改革的一个热门话题。实施双语教学是我国高等教育适应经济全球化趋势，培养具有国际合作意识、国际交流与竞争能力的外向型人才的重要途径。双语教学可分为三个层次，即全英语式教学、整合式双语教学和任务型语码转换式双语教学。前两个层次对师资和学生的外语水平要求较高，难以大面积实施，而任务型语码转换式双语教学则不受师资和学生的外语水平条件的限制，在教学中易于普及推广。

任务型语码转换式双语教学主要由两大模块构成：一、课堂上是在不影响授课进度和课堂信息量的前提下，根据学生实际、专业特点、学年变化及社会需求等，适时适量地渗透英文专业词汇、语句、语段或语篇，"润物细无声"地扩大学生的专业英文语汇，提高学生专业英语能力；二、课堂外向学生提供多种选择的"用中学"平台，如翻译英语科技文献，撰写英语实验报告、英语学术论文、英语小论文、英语课程设计报告等，使学生的专业英语实践应用达到一定频度和数量，激活英语与学科知识的相互渗透，培养学生用英语学习、科研、工作的能力及适应教育国际化、经济全球化的能力。

几年来，大连民族学院对任务型语码转换式双语教学进行了大量的研究与实践，经过广大教师的积极参与和共同努力，已在全校范围内取得了显著的成效，受到了学生的欢迎。该项研究在2004年荣获辽宁省第五届高等教育优秀教学成果一等奖，在2005年荣获国家级教学成果二等奖，得到了专家的肯定。为使这一成果得到推广和深化，便于广大教师在教学中实施渗透式双语教学，使学生从中实实在在地受益，我们组织我校机械专业承担主干课程的相关教师编写了这本面向机械类专业的任务型语码转换式双语教学辅助教材，希望对推动双语教学能有一定的帮助。

本书的编写有如下特点：(1)课程范围广。本书分为两篇：第一篇为基础篇，涵盖了机械类专业的技术基础课及主要专业课中的十八门课程，共分十八章，每章对应一门课程；第二篇为应用篇，主要包括摘要写作、实验设备简介、课程实验报告单、常用软件简介、面试情景对话等内容。(2)注重先进性。各门课程的内容依据近几年出版的优秀教材知识体系组织编写，英文词汇力求选自具有较大影响且为知名大学所采用的原版教材。(3)编排力求系统性。每门课程均按教材章节顺序给出专业词汇及短语，并选择适量的短句或语段译成中文。(4)使用便捷性。本书可用两种方法进行查阅：一种是按照课程的章节顺序进行查阅；另一种是按中文索引查找词条的出处，再查阅正文。

本书可作为机械类本科专业任务型语码转换式双语教学教材使用。教师在教学中可以从讲授课程对应的章节中选择适量的词汇或短语在教学过程中渗透给学生,以每节课3~5个单词为宜,日积月累,可丰富学生的专业英文词汇。讲授没有列入的课程时,可从中文索引或相关章节中选择与本课程相关的词汇进行渗透。学生可以把本书作为机械专业英文语汇的日常工具书,用于日常学习和查找专业词语。本书也可作为机械类专业教师、研究生及工程技术人员查找和丰富专业英文词汇的参考书使用。

参加编写的人员有管莉娜、魏莉(第一章);何韶君、白兰(第二、三、十六章);罗跃纲(第四章);冯长建(第五、十三章及应用篇"四、常用软件简介");吴斌、于善平(第六、七、八章及应用篇"二、实验设备简介");李文龙(第九、十章);胡红英(第十一、十二章);包耳、孙禹辉(第十四、十五章);邵强(第十七、十八章);康晶(应用篇"一、摘要写作");白兰(应用篇"三、大连民族学院课程实验报告单"和"五、面试情景对话"),梁艳君参加了部分章节中部分内容的编写与修改工作。

感谢周世宽老师做的大量编辑修改工作!在本书的编写过程中,查阅和参考了大量的文献资料,得到了许多有益的启发和教益,在此谨向参考文献的作者致以诚挚的谢意。

编写双语教学辅助教材是一个尝试,它不同于汉英词典,是按课程进行编排的。不论课程选择的合理性,还是教材选择的适当性,都有待于通过教学实践来检验。在本书编写过程中,虽然参考了国内外有关文献,但限于时间仓促,篇幅有限,很难全面反映机械领域中各学科的内容,存在错误和不足之处在所难免,希望读者批评指正。

编 者

2014年5月

使用说明

  本书主要是面向机械类专业的教师和学生编写的,适用于机械类专业的任务型语码转换式双语教学。书中选择了机械专业的技术基础课及主要专业课中的 18 门课程,每章对应一门课程,每节为课程的一个知识单元。每节中先给出相关的单词和短语,然后是短句或语段。

  本书可从两方面进行查阅。一种是按照课程的章节顺序进行查阅,即在词汇所属课程相应的章节中查找相关的词汇或短语。另一种是按索引法,即按照英文字母顺序词条索引查找词条的出处,再查阅正文。例如,要查找"齿轮"一词,从中文索引中可查到"齿轮 10-11","10-11"即指该词条在第十章第十一节中出现,从正文中即可查到该词条的英文为"gear"。

第一部分 基础篇 /1

第一章 画法几何与机械制图 /1
- 第一节 制图基本知识和技能 /1
- 第二节 投影基础 /1
- 第三节 立体的投影 /2
- 第四节 组合体 /2
- 第五节 轴测投影 /2
- 第六节 机件的表达方法 /3
- 第七节 标准件和常用件 /3
- 第八节 零件图 /4
- 第九节 装配图 /4
- 第十节 其他工程图介绍 /5

第二章 机械原理 /6
- 第一节 机构的结构分析 /6
- 第二节 机构的运动分析 /6
- 第三节 机构的动力学分析 /7
- 第四节 连杆机构 /8
- 第五节 凸轮机构 /9
- 第六节 齿轮机构 /9
- 第七节 其他机构 /10

第三章 机械设计 /11
- 第一节 总论 /11
- 第二节 连接 /11
- 第三节 机械传动 /12
- 第四节 轴系零部件 /13
- 第五节 其他零部件 /14

第四章 材料力学 /16
- 第一节 绪论 /16
- 第二节 轴向拉伸和压缩 /16
- 第三节 扭转 /18
- 第四节 弯曲内力 /18
- 第五节 弯曲应力 /19
- 第六节 弯曲变形 /19
- 第七节 应力状态和强度理论 /20
- 第八节 组合变形 /21
- 第九节 压杆稳定性 /22
- 第十节 平面图形的几何性质 /22

第五章 理论力学 /24
- 第一节 绪论 /24
- 第二节 静力学的基本原理 /24
- 第三节 平面特殊力系 /25
- 第四节 一般力系 /26
- 第五节 质点的运动 /26
- 第六节 刚体的基本运动 /27
- 第七节 质点运动的合成 /27
- 第八节 刚体的平面运动 /28
- 第九节 刚体的一般运动 /28
- 第十节 质点运动微分方程 /28
- 第十一节 动量定理 /29
- 第十二节 动量矩定理 /30
- 第十三节 动能定理 /31
- 第十四节 达朗伯原理 /31

第六章 机械制造技术基础及装备设计 /33
- 第一节 机械加工方法 /33
- 第二节 金属切削原理与刀具 /33
- 第三节 金属切削机床及设计 /35
- 第四节 机床夹具原理与设计 /37
- 第五节 加工质量分析与控制 /38
- 第六节 工艺规程设计 /39
- 第七节 物料输送系统及仓储装置设计 /39
- 第八节 机械加工生产线 /39
- 第九节 先进制造技术 /40

第七章 微机原理与应用 /41
- 第一节 计算机基础知识 /41
- 第二节 MCS-51 单片机概述 /41
- 第三节 单片机的结构原理 /42
- 第四节 MCS-51 指令系统 /42
- 第五节 程序设计 /43
- 第六节 硬件基础 /44

第八章 冷冲压模具设计 / 45
第一节 冲裁模具设计基础 / 45
第二节 冲裁工艺与冲裁模设计 / 45
第三节 弯曲工艺与弯曲模具设计 / 45
第四节 拉深工艺与拉深模具设计 / 46
第五节 模具加工方法与刀具 / 46
第六节 模具加工机床与夹具 / 47

第九章 计算机辅助绘图 / 48
第一节 AutoCAD 概述 / 48
第二节 绘图命令栏 / 48
第三节 编辑命令栏 / 49
第四节 辅助绘图命令 / 49
第五节 标注命令 / 50
第六节 综合运用各种绘图方法绘制工程图 / 50
第七节 Pro/E / 51
第八节 SolidWorks 的基本功能 / 51

第十章 互换性与技术测量 / 53
第一节 圆柱公差与配合 / 53
第二节 长度测量基础 / 53
第三节 形状和位置公差 / 55
第四节 表面粗糙度及检测 / 55
第五节 光滑极限量规 / 55
第六节 滚动轴承的公差与配合 / 56
第七节 尺寸链 / 56
第八节 圆锥的公差配合及检验 / 56
第九节 螺纹公差及检测 / 56
第十节 键和花键的公差与配合 / 57
第十一节 圆柱齿轮传动公差及检测 / 57

第十一章 计算机辅助设计/计算机辅助制造技术 / 58
第一节 概述 / 58
第二节 CAD/CAM 系统的硬件和软件 / 58
第三节 CAD/CAM 系统的开发基础 / 59
第四节 计算机图形学 / 59
第五节 实体建模 / 60
第六节 计算机辅助工程 / 60
第七节 计算机辅助工艺规程设计 / 61
第八节 CAD/CAM 集成和计算机集成制造系统 / 61

第十二章 非传统加工/特种加工 / 63
第一节 绪论 / 63
第二节 电火花加工 / 63
第三节 电火花线切割 / 63
第四节 电化学加工 / 64
第五节 激光束加工 / 64
第六节 电子束和离子束加工 / 64
第七节 超声加工 / 65
第八节 快速成型技术 / 65
第九节 特种加工方法 / 66

第十三章 塑料成型与模具制造 / 67
第一节 注射模具 / 67
第二节 重叠成型 / 73
第三节 气体辅助注射成型 / 74
第四节 共注射成型 / 75
第五节 注射压缩成型工艺 / 76
第六节 反应成型工艺 / 76
第七节 词汇表 / 78

第十四章 机械工程材料 / 91
第一节 工程材料的性能 / 91
第二节 材料结构 / 91
第三节 材料的凝固 / 92
第四节 二元相图及其应用 / 92
第五节 材料的变形 / 93
第六节 钢的热处理 / 93
第七节 工业用钢 / 94
第八节 铸铁 / 95
第九节 有色金属及其合金 / 95
第十节 常用非金属材料 / 96
第十一节 新型材料 / 96
第十二节 工程材料的选用 / 97

>> 第十五章 金属工艺学 / 98
第一节 金属材料基本知识 / 98
第二节 铸造 / 98
第三节 压力加工 / 99
第四节 焊接 / 100
第五节 切削加工 / 101

>> 第十六章 机械设计课程设计 / 102

>> 第十七章 控制工程 / 104
第一节 绪论 / 104
第二节 系统的数学模型 / 104
第三节 系统的时间响应分析 / 105
第四节 系统的频率特性分析 / 105
第五节 系统的稳定性 / 106
第六节 系统的性能指标与校正 / 106

>> 第十八章 机电测试技术 / 108
第一节 绪论 / 108
第二节 信号及其描述 / 108
第三节 信号及其描述 / 109
第四节 传感器 / 109
第五节 信号调整、处理 / 110
第六节 信号的记录 / 111

>> 第二部分 应用篇 / 113

>> 一、摘要写作 / 113

>> 二、实验设备介绍 / 115

>> 三、大连民族学院课程实验报告单 / 118

>> 四、常用软件介绍 / 120

>> 五、面试情景对话 / 137

>> 参考文献 / 140

>> 索 引 / 141

第一部分　基础篇

第一章　画法几何与机械制图
Chapter 1　Descriptive Geometry and Mechanical Drawing

第一节　制图基本知识和技能
Section 1　The Fundamental Knowledge and Skill of Drawing

- ★ 国家标准　national standard (GB)
- ★ 机械制图　mechanical drawing
- ★ 工程图学　engineering graphics
- 计算机绘图　computer graphics
- ★ 计算机辅助设计　computer added design
- ★ 图纸幅面　formats of drawing
- 图框　border of drawing
- ★ 标题栏　title block
- ★ 尺寸标注　dimensioning
- ★ 几何作图　geometric drawing
- ★ 几何图形　geometrical figure
- ★ 圆弧连接　arc connection
- ★ 平面图形　plane figure
- ★ 定形尺寸　shape dimension
- ★ 定位尺寸　location dimension
- ★ 线段分析　line segment analysis
- ★ 绘图步骤　drawing order
- ★ 绘图仪器　drawing instrument
- ★ 图板　drawing board
- ★ 丁字尺　T-square
- ★ 三角板　set square/triangle
- 擦图片　erasing shield
- 透明胶带　scotch tape/cellulose tape
- 细砂纸　sandpaper
- 量角器　protractor
- 曲线板　french curve
- 模板　template
- ★ 徒手绘图　freehand drawing
- 常用制图工具和仪器　instruments and materials in common use
- 制图基本规格　general standards of drawing
- 图线及其画法　basic conventions for lines

第二节　投影基础
Section 2　The Fundamentals of Projection

- 投影图　projection drawing
- ★ 实形性　characteristic of true
- ★ 积聚性　characteristic of concentration
- ★ 类似性　characteristic of similarity
- ★ 三视图　three-view drawing
- 投影基本知识　fundamental knowledge of the projection
- 点的三面投影　point of projection on three plans
- 直线的投影特性　characteristic on projection of line
- 直线的实长和倾角　true length and dip of a line
- 重影　coincident projection
- 交点　point of intersection
- ★ 投影面平行面　plane parallel to the projection plane
- 斜线　inclined line
- 交叉线　skew line
- 直角投影定理　theorem of right angle projection
- ★ 投影面垂直面　plane perpendicular to the projection plane
- ★ 一般位置直面　oblique plane
- 平面内的点和直线　point and line on a plane
- 平行关系　parallel relation
- 相交关系　intersection relation
- 垂直关系　perpendicular relation
- ★ 主视图　front view
- ★ 俯视图　top view
- ★ 左视图　left view

★ 换面法 transformation plane method
★ 中心投影法 central projection
★ 平行投影法 parallel projection
★ 正投影法 orthographic projection
  斜投影法 oblique projection
  投影特性 characteristic of projection

第三节 立体的投影
Section 3 The Projection of Geometrical Bodies

★ 平面立体 plane solid
基本体的投影 projection of basic geometrical body
立体及其表面的点 solid and point on its surface
★ 正六棱柱 right hexagonal prism
★ 三棱锥 triangular pyramid
  正四棱锥 right square pyramid
  顶点 center vertex
★ 棱锥台 truncated pyramid/frustum of pyramid
  曲面立体 body of curved surface
  回转体 rotative body
  转向线 change direction outline
  几何元素 geometric element
  对称线 symmetrical line
  辅助线 auxiliary line
  平面与立体相交 intersections of plane and solid
★ 截交线概念和性质 conception and character of cut line
  立体与立体相交 intersections of solid and solid
★ 相贯线的概念和性质 conception and character of intersecting line
  组合相贯线 complex intersecting line
  相贯体 intersecting body
★ 辅助平面法 auxiliary plane method
★ 圆锥顶 conic apex
  回转体 intersection of revolution solid

第四节 组合体
Section 4 Combination Solids

★ 组合体 complex/combination solid
  组合体的三视图 three-views of a complex
★ 组合体视图的画法 drawing view of a complex
★ 组合体的形成方式 formation system of a complex
  组合体的分析 analysis of a complex
★ 组合体的尺寸标注 dimensioning of a complex
  标注尺寸的基本要求 basic requirement of dimensioning
★ 读组合体的视图 read the views of a complex
  读图的基本要领 fundamental key points of reading drawings
  几何形体 geometry bold
  叠加 piling up/built-up
★ 切割 cutting
  贯穿 penetration
  支架 brace
  底板 bottom board
  肋 rib
  支撑板 bearing plate
★ 形体分析 shape-body analysis
★ 线面分析 line-surface analysis
  基准线 datum line
  基准面 datum plane
  中心孔 central hole
  中心距 center distance
  表面连接关系 surface connecting relationship
★ 立体构型 spatial configuration

第五节 轴测投影
Section 5 Axonometric Projection

★ 轴测图 axonometric drawing
  轴向伸缩系数 axial deformation coefficient
  轴间角 axes angle
  轴测图的分类 classification of axonometric projection
★ 正等轴测图 right isometric projection/isometric projection
★ 斜二等轴测 oblique dimetric projection
  轴测投影面 plane of axonometric projection
  轴测剖视图 isometric sectional view
  缩短 foreshortened
★ 切割法 cutting method/by cutting

- ★ 方箱法　boxing method
- ★ 坐标法　coordinate method
- 方框法　enclosing-square method
- 四心近似法　four-center approximate method

第六节　机件的表达方法
Section 6　The Representation of Parts

- ★ 基本视图　basic view
- ★ 右视图　right view
- 仰视图　bottom view
- 后视图　rear view
- 局部视图　partial view/broken view
- 斜视图　oblique view
- ★ 向视图　reference arrow layout view/direction drawing
- ★ 剖视图　sectional view/section
- 剖视图的概念　conception of section
- 剖视图的类型　classification of sections
- ★ 全剖视图　full section
- ★ 半剖视图　half section
- ★ 局部剖视图　partial section/broken section
- ★ 斜剖视图　oblique section
- 旋转剖视图　aligned section
- ★ 阶梯剖视图　offset section
- 复合剖视图　compound section
- ★ 断面图　cut section/section
- 移出断面图　removed section
- 重合断面图　coincidental section/revolved section
- 断裂边界　broken boundary
- ★ 局部放大图　partial enlarged view
- ★ 简化画法　simplified representation
- ★ 规定画法　conventional representation
- 省略画法　omissive representation
- 辅助图　auxiliary view

第七节　标准件和常用件
Section 7　Standard Parts and Commonly Used Parts

- 规定画法　stipulation representation
- 特殊表示法　special representation
- 螺纹的基本要素　basic essential factors of thread
- ★ 螺纹牙型　screw thread profile
- 常用螺纹的种类和标注　kind and symbol of thread in common use
- ★ 粗牙普通螺纹　coarse metric thread
- ★ 细牙普通螺纹　fine metric thread
- ★ 圆柱管螺纹　cylindrical pipe thread
- ★ 圆锥管螺纹　conical pipe thread
- ★ 梯形螺纹　trapezoidal thread
- ★ 锯齿形螺纹　buttress thread
- ★ 公称直径　nominal diameter
- ★ 螺距　pitch of thread/pitch
- ★ 导程　lead of thread/lead
- ★ 螺纹线数　number of thread
- ★ 旋向　direction of turning
- ★ 螺纹紧固件的标记　symbol of threaded fastener
- 螺纹紧固件的连接形式　joint forms of threaded fastener
- ★ 六角头螺栓　hexagon head bolt
- ★ 双头螺柱　double end studs
- ★ 六角螺母　hexagon nut
- 螺栓连接　bolt joint/fastening
- 双头螺柱连接　stud joint/stud fastening
- 螺钉连接　screw joint
- ★ 普通平键　flat key
- 半圆键　half round key
- 钩头楔键　wedge key/gid head key
- ★ 圆柱销　dowel pin
- ★ 圆锥销　taper pin
- ★ 圆柱齿轮　cylindrical gear
- 斜齿轮　spiral gear
- ★ 正齿轮　spur gear
- 蜗轮与蜗杆　worm wheel and worm gear
- ★ 分度圆　reference circle
- ★ 模数　module
- 压力角　pressure angle
- 锥齿轮　bevel gear
- ★ 蜗轮　worm wheel
- 滚动轴承的代号　code name of rolling bearing
- 向心轴承　radial bearing
- 密封装置　sealing equipment
- 锁紧装置　locking equipment

第八节 零件图
Section 8　Detail Drawing

★ 零件图的作用和内容　action and content of detail drawing
零件的形状结构　shape structure of detail
典型零件的视图选择　view choosing of typical detail
★ 轴套类零件　axle-sleeve detail/shaft-sleeve parts
★ 盘盖类零件　disk-cover detail/disk-shaped parts
★ 叉架类零件　fork-rack detail/fork-shaped parts
★ 箱体类零件　box detail/case-shaped parts
★ 设计基准　design datum
★ 工艺基准　technology datum
★ 零件的技术要求　technique requirement of detail
合理标注尺寸的方法　appropriate dimensioning method
工艺结构　technological feature/technological structure
加工位置原则　machining position principle
工作位置原则　functional position principle
形状特征原则　shape characteristic principle
★ 表面粗糙度　surface roughness
★ 极限与配合　limits and fits
★ 互换性　interchangeability
★ 尺寸公差　size tolerance
尺寸公差带　size tolerance zone
★ 标准公差　standard tolerance (IT)
★ 基本偏差　fundamental deviation
标准公差等级　standard tolerance grade
间隙配合　clearance fit
过盈配合　interference fit
过渡配合　transition fit
基孔制　hole-basis system
基轴制　shaft-basis system
标注和查表　symbol and lookup table
形状和位置公差　geometrical tolerance
零件结构的工艺性　technological property of detail structure
铸造零件的工艺结构　technological structure of cast detail
机械加工工艺结构　technological structure of machine process
铸造工艺结构　technological structure of casting process
锪平沉孔　sport face
直纹滚花　straight knurling
网纹滚花　hatching knurling
★ 零件测绘　surveying and drawing of part

第九节 装配图
Section 9　Assembly Drawing

★ 零件序号　serial number of detail
★ 明细栏　item list
装配结构的合理性　reasonableness of assemble structure
接触面与配合面的合理结构　reasonable structure of contact surface and fit surface
装配关系和工作原理　mounting relation and working principle
★ 表达方案　representation scheme
★ 由装配图拆画零件图　dismantle assembly drawing to detail drawing
★ 拆卸画法　taking some parts apart
假想画法　representation of using phantom line
特殊画法　special representation
夸大画法　exaggerated representation
★ 规格尺寸　characteristic dimension
★ 装配尺寸　assembly dimension
★ 安装尺寸　installation dimension
★ 总体尺寸　outer dimension/overall dimension
★ 配合尺寸　fit dimension
装配工艺结构　technological feature of assembly processes
零部件测绘　survey and drawing on detail and assembly
拆卸零件　dismantle details
装配示意图　assembly diagram

第十节　其他工程图介绍
Section 10　Other Engineering Drawings

展开图　developed drawing
立体表面的展开　development of solid surface
可展曲面　developable curved surface
近似展开　approximate development
螺旋面　helical convolute

焊接结构　welding construction
焊接图　welding drawing
焊缝　welding seam
焊缝符号　welding symbolic
焊缝接头　welding joint

第二章 机械原理
Chapter 2 Machine Principles

第一节 机构的结构分析
Section 1 Structure Analysis of Mechanism

机械原理	mechanism theory	圆柱副	cylindric pair
机械学	mechanics	球面副	globular pair
机器	machine	转动副	rotating pair
机械	machinery	移动副	sliding pair
机构	mechanism	螺旋副	screw pair
零件	component	闭式链	closed chain
构件	member	开式链	open chain
机架	frame	运动简图	kinematic scheme
结构分析	structure analysis	自由度数	number of degree of freedom
运动副	kinematic pair	复合铰链	multiple pin joint
运动副元素	element of kinematic pair	局部自由度	partial freedom
高副	higher pair	虚约束	formal constraint
低副	lower pair	运动确定性	kinematic determination
平面副	flat pair		

1 Kinematic pair is a connection between two or more links (at their nodes), which allows some motion, or potential motion between the connected links.
运动副是一种在两个或更多构件接触点处的连接方式,构件间应有相对运动。

2 A kinematic chain is a system of links, that is, rigid bodies, which are either jointed together or are in contact with one another in a manner that permits them to move relative to one another. If one of the links is fixed and the movement of any other link to a new position will not cause each of the other links to move to definite predictable position, the system is a constrained kinematic chain. If one of the links is held fixed and the movement of any other link to a new position will not cause each of the other links to move to a definite predictable position, then the system is an unconstrained kinematic chain.
运动链是一个构件系统即若干个刚体,它们或者彼此铰接或者相互接触,并允许彼此间产生相对运动。如果构件中的某一构件被固定,而任何其他一个构件运动到新的位置将会引起其他各个构件也运动到确定的预期位置,该系统就是一个可约束的运动链。如果构件中的某一个构件保持固定而任一构件运动到一新的位置不会使其他各个构件运动到一个确定的预期位置,则该系统是一个非约束运动链。

第二节 机构的运动分析
Section 2 Kinematic Analysis of Mechanism

主动构件	driving member	瞬时速度	instantaneous velocity
从动构件	driven member	相对速度	relative velocity
速度极点	velocity pole	加速度图	acceleration diagram
速比	velocity ratio	加速度极点	acceleration pole
绝对速度	absolute velocity	角加速度	angular acceleration
等角速度	constant angular velocity	哥氏加速度	coriolis acceleration

法向加速度　normal acceleration	速度影像　velocity image
速度瞬心　instantaneous center of velocity	矢量方程式　vector equation
绝对瞬心　absolute instantaneous center	矢量多边形　vector polygon
相对瞬心　relative instantaneous center	速度多边形　velocity polygon
瞬心位置　location of instantaneous center	加速度多边形　acceleration polygon
影像法　image construction	

1 The definition of an instantaneous center of velocity is a point, common to two bodies in plane motion, which has the same instantaneous velocity in each body.

速度瞬心是指两个构件互作平面相对运动时,被认为是在绕某一点做相对转动,且在该点两构件的瞬时速度相同,则这一点称为速度瞬心。

2 Another method is the instantaneous center or instant center method, which is a very useful and often quicker in complex linkage analysis. An instantaneous center or instant center is a point at which there is no relative velocity between two links of a mechanism at that instant. In order to locate the locations of some instant center of a given mechanism, the Kennedy's theorem of three centers is very useful. It states that the three instantaneous centers of three bodies moving relative to one another must lie along a straight line.

另一种方法是瞬时中心法,即瞬心法。该方法非常有用,是复杂连杆机构分析时速度较快的方法。瞬心是一个点,该点在那一瞬间,机构上的两构件之间不存在相对运动。为了找出一支机构某些瞬心的位置,肯尼迪(Kennedy)三中心理论非常有用。它是说:彼此相对运动的三个物体的三个瞬心必定是在一直线上。

第三节　机构的动力学分析
Section 3　Dynamics Analysis of Mechanism

机械的动能方程式　kinetic energy equation of machinery	自锁　self locking
有效力　effective force	平衡　balance
等效力　generalized force	配重　counter balance
等效运动方程式　generalized equation of motion	动平衡　dynamic balance
等效构件　generalized link	静平衡　static balance
等效质量　generalized mass	平衡试验　balancing test
等效力矩　generalized torque	平衡重量　balancing weight
等效转动惯量　generalized moment of inertia	平衡精度　accuracy of balance
摩擦圆　friction circle	质径积　mass-distance product
当量摩擦系数　equivalent friction factor	调节　adjustment
当量摩擦角　equivalent friction angle	可调节性　adjustability
惯性力　inertia force	能量守恒　conservation of energy
惯性力矩　inertia torque	盈亏功　fluctuation of energy
机组　mechanical system	调速器　governor
反行程　reverse drive	离心调速器　centrifugal governor

1 Kinematics: the study of motion without regard to forces. Kinetics: the study of forces on systems in motion. These two concepts are really not physically separable. We arbitrarily separate them for instructional reasons in engineering education. It is also valid in engineering design practice to first consider the desired kinematic motions and their consequences, and then subsequently investigate the kinetic forces associated with those motions. Students should realize that the division between kinematics and kinetics is quite arbitrary and is done largely for convenience. One cannot design most dynamic mechanical systems without taking both topics into thorough consideration. It is quite logical to consider them in the order listed since, from Newton's second law, $F = ma$, one typically needs to know the accelerations (a) in order to compute the dynamic forces (F) due to the motion of the system's mass (m).

运动学:不考虑力的运动的研究。动力学:运动系统中对力的研究。这两种概念并不是完全分离的,我们区别它们是为了工程教学的需要。在工程设计实践中首先确定预期的运动学中运动的结果,然后研究与运动有关系的力。学生们要认识到运动学和动力学的区别是人为的,并且主要是为了方便学习。任何人都不能设计一个机械运动系统而能不包括这两种概念。从列举牛顿第二定律可以做出合理的解释,根据牛顿第二定律,$F = ma$,为了计算力人们通常要知道加速度和运动系统的质量。

2 Simply speaking, the requirement for static balance is that the sum of all forces on the moving system (including d'Alembert inertial forces) must be zero.

简而言之,静平衡是要求运动系统中所有惯性力之和必须为零。

第四节 连杆机构
Section 4　Linkages Mechanism

曲柄摇杆机构　crank-and-rocker mechanism	双摇杆机构　double-rocker mechanism
曲柄摇块机构　crank-swing block mechanism	摇块机构　rocking-block mechanism
双曲柄机构　double-crank mechanism	正弦机构　sine mechanism
曲柄导杆机构　crank-and-guide bar mechanism	曲柄滑块机构　slider-crank mechanism
曲柄摆动导杆机构　crank-and-oscillating guide-bar mechanism	滑块的导路　slider guide
	对心曲柄滑块机构　central(radial) slider-crank mechanism
曲柄转动导杆机构　crank-and-rotating guidebar mechanism	十字滑块机构　crossed-slider mechanism
曲柄移动导杆机构　crank-and-translation guide-bar mechanism	双滑块机构　double-slider mechanism
	偏置曲柄滑块机构　eccentric slider-crank mechanism
铰链四杆机构　four-bar linkage mechanism	
平行四边形机构　parallelogram linkage mechanism	行程速度变化系数　coefficient of travel speed variation
平面连杆机构　planar linkage mechanism	曲柄存在条件　conditions of crank existence
空间连杆机构　spatial linkage mechanism	连杆曲线　coupler curve
曲柄摇杆机构　quadric-crank mechanism	传动角　transmission angle
急回运动　quick-return motion	传动特性　transmission characteristic
急回特性机构　quick-return mechanism	死点　dead point

1 The transmission angle is defined as the angle between the output link and the coupler. It is usually taken as the absolute value of the acute angle of the pair of angles at the intersection of the two links and it varies continuously from some minimum to some maximum value as the linkage goes through its range of motion.

传动角是指输出杆和连杆间的夹角,它的值等于输出杆与连杆间所夹锐角,或等于所夹钝角的补角的值。它随机构的运动而逐渐变大。

2 Path generation. A path generation mechanism will guide a point on a rigid body through a series of points on a specified path in space. Function generation. A mechanism that creates an output motion that is a specified function of the input motion.

实现轨迹。实现轨迹机构将引导刚体上的一个点,使其通过指定的空间轨迹上的一系列点。实现函数。这类机构所产生的输出运动是输入运动的指定函数。

第二章 机械原理

第五节 凸轮机构
Section 5　Cams Mechanism

刚性冲击　rigid impulse
柔性冲击　soft impulse
等速运动规律　law of uniform motion
等加速等减速运动规律　law of constant acceleration and deceleration motion
余弦加速度运动规律　law of cosine acceleration motion
正弦加速度运动规律　law of sine acceleration motion
盘形凸轮　disk cam/radial cam
圆柱凸轮　barrel cam/cylindrical cam
圆锥形凸轮　conical cam
球面凸轮　spherical cam
运动失真　distortion of motion
从动件滚子　follower roller
摆动从动件　pivoted follower
端从动件　tip follower
滚子从动件　roller follower
平底从动件　plain-faced follower
移动从动件　translating follower
凸轮轮廓　cam profile
凸轮轮廓上的尖点　sharp point at the cam profile
凸轮理论轮廓线　theoretical cam profile
基圆　base circle
升程　advanced travel

1 The cam is projection on a wheel or shaft, designed to change circular motion into up-and-down or back-and-forth motion.
凸轮是轮子或轴上的凸出物,它被用来使圆周运动转变为上下或前后运动。

2 A cam is a convenient device for transforming one motion into another. This machine element has a curved or grooved surface which mates with a follower and imparts motion to it. The motion of the cam (usually rotation) is transformed into follower oscillation, translation or both. Because of the various cam geometries and the large number of cam and follower combinations, the cam is an extremely versatile mechanical element. Although a cam and follower may be designed for motion, path generation, the majority of applications utilize the cam and follower for function generation.
凸轮装置是把一种运动改变成另一种运动的简便装置。这种机器零件有曲面或斜槽面,该曲面或斜槽面与从动件相配合并将运动传递给从动件。凸轮的运动(通常是转动)使从动件作摆动或移动,或两者均有。由于凸轮可以采用各种各样的几何形状并与从动件有各种各样的组合方式,因此凸轮是一种非常通用的机械零件。虽然凸轮和从动件可以传递运动或轨迹,但主要还是用于实现某种功能。

第六节 齿轮机构
Section 6　Gear Mechanism

渐开线齿轮　involute gear
摆线齿轮　cycloidal gear
标准齿轮　standard gear
变位齿轮　modified gear
螺旋齿轮　spiral gear
人字齿轮　double helical gear
直齿圆锥齿轮　straight bevel gear
当量齿轮　virtual gear
渐开线函数　involute function(Inv)
啮合点　meshing point
实际啮合线长度　working length of meshing line
标准压力角　standard pressure angle
蜗杆　worm/worm screw
蜗轮　worm gear/worm wheel
阿基米德蜗杆　Archimedes worm
啮合角　angle of action
齿廓啮合基本定律　fundamental law of gear-tooth action
啮合线　line of action
齿顶圆直径　diameter of addendum
基圆直径　base diameter
公法线长度　base tangent length
齿厚　circular thickness
极限啮合点　limiting contact point
工作齿廓　working tooth contour
惰轮　idle gear
差动轮系　differential gear train
周转轮系　epicyclic gear train

混合轮系　compound epicyclic gear train
定轴轮系　ordinary gear train
齿轮系　train of gear
行星转臂　planet cage
行星齿轮　planetary gear
行星轮系　planetary gearing/planetary gear train
单级行星轮系　single planetary gear train
多级行星轮系　compound planetary gear train
谐波齿轮传动　harmonic gear transmission

1 Spur gears are the most widely used style of gears and are used to transmit rotary motion between parallel shafts, while maintaining uniform speed and torque. The involute tooth form, being the simplest to generate, permits high manufacturing tolerances to be attained.

直齿轮是齿轮中应用最广泛的一种类型，它用在平行轴间传递旋转运动，并且保持恒定的速度和转矩。这种轮齿的渐开线齿廓是最易于生产的，并可获得很高的制造精度。

2 Epicyclic gear trains, if properly designed, can have even higher overall efficiencies than conventional trains. But, if the epicyclic gear train is poorly designed, its efficiency can be so low that it will generate excessive heat and may even be unable to operate at all. This strange result can come about if the orbiting elements (planets) in the train have high losses that absorb a large amount of "circulating power" within the train. It is possible for this circulating power to be much larger than the throughput power for which the train was designed, resulting in excessive heating or stalling. The computation of the overall efficiency of an epicyclic gear train is much more complicated than the simple multiplication indicated above that works for ordinary gear trains.

周转轮系，如果设计得当，可以得到比一般轮系更高的总效率。但是，如果周转轮系设计不得当，它的效率可能会很低以至于它会产生过度的热，甚至根本无法运转。这个特殊结果表明，如果行星轮系有很高的损耗以至于在轮系中吸收了大量的循环功率，这个循环功率比设计轮系时的生产功率要高很多，结果是轮系产生过度的热量和停车。周转轮系总效率的计算要比定轴轮系的简单乘法复杂得多。

第七节　其他机构
Section 7　Miscellaneous Mechanism

棘轮机构　ratchet mechanism
外棘轮机构　external ratchet mechanism
内棘轮机构　internal ratchet mechanism
间歇机构　dwell mechanism
槽轮机构　Geneva mechanism
外槽轮机构　external Geneva mechanism
内槽轮机构　internal Geneva mechanism
擒纵机构　escapement mechanism
万向联轴节　universal joint
单万向联轴节　single universal joint
双万向联轴节　double universal joint
等速万向联轴节　constant velocity universal joint
齿轮连杆机构　geared linkage mechanism

The ratchet wheel is the toothed wheel provided with a catch that prevents the wheel from slipping back and allows it to move in only one direction.

棘轮是指啮合起来的轮子，且带有一个钩状抓取物，能防止轮子倒滑，并能使轮子只能朝一个方向运动。

第三章 机械设计
Chapter 3 Design of Machinery

第一节 总论
Section 1 Introduction

无限寿命设计	infinite life design	耐磨性	wear resistance
有限寿命设计	finite life design	液体润滑	fluid lubrication
接触疲劳	contact fatigue	固体润滑	solid-film lubrication
循环载荷	cyclic loading	液体静力润滑	hydrostatic lubrication
随机载荷	random loading	弹性流体动力润滑	elasto-hydrodynamic lubrication
循环特性	stress ratio		
脉动循环	pulsation cycle	连续润滑	continuous lubrication
对称循环	symmetry cycle	间歇润滑	periodical lubrication
疲劳强度	fatigue strength	循环润滑	circulating lubrication
疲劳极限	fatigue limit	油浴润滑	bath lubrication
疲劳积累损伤	cumulative fatigue damage	油雾润滑	mist lubrication
流体摩擦	fluid friction	油环润滑	oil-ring lubrication
边界摩擦	boundary friction	润滑剂	lubricant
磨料磨损	abrasive wear	添加剂	additive
剥落	spalling	可靠性	reliability
接触疲劳	contact fatigue	失效	failure
点蚀	pitting	磨损	wear
磨损率	wear rate	故障	fault

1 Design of machine elements is an integral part of the larger and more general field of mechanical design. In this book, you will be introduced to learn the principles of machine elements in mechanical design.

在大型的、综合的机械设计领域中,机械零件设计是其中一个完整的体系。在这本书中,将学到机械设计中的机械零件理论。

2 The definition of friction is the rubbing of one thing against another, esp., when this wastes energy.

摩擦是指某一构件对于另一构件的运动阻碍过程,同时损耗能量。

3 The general definition of efficiency is output power/input power expressed as a percentage. A spur gearset can be 98% to 99% efficiency.

效率的一般定义是输出功率与输入功率之比。直齿轮机构的效率可以达到98%~99%。

第二节 连接
Section 2 Joint

圆柱螺纹	parallel screw thread	螺纹大径	major diameter
圆锥螺纹	taper screw thread	螺纹小径	minor diameter
外螺纹	external thread	螺纹中径	pitch diameter
内螺纹	internal thread	螺距	pitch
牙型角	thread angle	螺纹导程	lead
公称直径	nominal diameter	螺纹升角	lead angle

双头螺柱　stud	键槽　key way
六角头螺栓　hexagon bolt	普通平键　general flat key
地脚螺栓　foundation bolt	导向平键　dive key
紧定螺钉　set screw	滑键　feather key
自攻螺钉　tapping screw	半圆键　woodruff key
吊环螺钉　lifting eye bolt	楔键　taper key
内六角圆柱头螺钉　hexagon socket head cap head screw	钩头楔键　gib-head taper key
	切向键　tangential key
六角螺母　hexagon nut	矩形花键　rectangle spline
圆螺母　round nut	渐开线花键　involute spline
垫圈　washer	销　pin
平垫圈　plain washer	圆柱销　cylindrical pin/straight pin
弹性垫圈　spring washer	圆锥销　conical pin/taper pin
键　key	槽销　grooved pin
花键　spline	安全销　safety pin

Typical methods of fastening and joining parts include the use of such items as bolts, nuts, cap screws, setscrews, rivets, locking devices and keys. Parts may also be joined by welding, brazing, or clipping together. Studies in engineering graphics and in metal processes often include instruction on various joining methods, and the curiosity of any person interested in engineering naturally results in acquisition of a good background of fastening methods.

典型的紧固和连接零件的方法包括螺栓、螺帽、有头螺钉、定位螺钉、铆钉、锁紧装置和键等。零件也可以用熔焊、铜焊和夹紧连接。在工程图学和金属加工工艺研究中常常涉及关于连接方法的说明。在工程上对此很感兴趣的人或求知欲强的任何人自然会获得关于紧固方法的基本知识。

第三节　机械传动
Section 3　Mechanical Drive

带传动　belt drive	联组 V 带　joined V-belt
平带传动　flat belt drive	多楔带　poly V-belt
V 带传动　V-belt drive	基准宽度　datum width
圆带传动　round belt drive	基准直径　datum diameter
同步带传动　synchronous belt drive	节宽　pitch width
主动带轮　driving pulley	链传动　chain transmission
从动带轮　driven pulley	链条　chain
带长　belt length	滚子链　roller chain
包角　angle of contact	单排滚子链　simplex roller chain
滑动率　sliding speed	双排滚子链　duplex roller chain
初拉力　initial tension	多排滚子链　multiplex roller chain
紧边拉力　tight side tension	传动链　transmission chain
松边拉力　slack side tension	输送链　conveyor chain
有效拉力　effective tension	内链节　inner link
离心拉力　centrifugal tension	外链节　outer link
普通平带　conventional belt	连接链节　connecting link
普通 V 带　conventional V-belt	链板　link plate
窄 V 带　narrow V-belt	链轮　chainwheel/sprocket
宽 V 带　wide V-belt	齿轮传动　gear drive

第三章 机械设计

平行轴齿轮传动　gear drive with parallel axes
相交轴齿轮传动　gear drive with intersecting axes
交错轴齿轮传动　gear drive with non-parallel non-intersecting axes
主动齿轮　driving gear
从动齿轮　driven gear
齿轮承载能力　load capacity of gears
圆柱齿轮　cylindrical gear
渐开线圆柱齿轮　involute cylindrical gear
圆弧圆柱齿轮　circular-arc gear
单圆弧齿轮　single-circular-arc gear
双圆弧齿轮　double-circular-arc gear
圆柱蜗杆　cylindrical worm
环面蜗杆　toroid worm/enveloping worm
阿基米德蜗杆　straight sided axial worm
渐开线蜗杆　involute helicoid worm
圆弧圆柱蜗杆　hollow flank worm
中间平面　mid plane
轴交角　shaft angle
蜗轮节圆　pitch circle of worm wheel
蜗杆头数　number of threads of worm
蜗杆旋向　hands of worm

1 The gear set is the toothed wheels working together in a machine. For example, a set to connect a motor-car engine with the road wheels.
齿轮机构是指相互啮合的轮子，它们在机器中同时工作。如将汽车发动机和车轮连接在一起的装置。

2 A gear having tooth elements that are straight and parallel to its axis is known as a spur gear. A spur pair can be used to connect parallel shafts only. Parallel shafts, however, can also be connected by gears of another type, and a spur gear can be mated with a gear of a different type.
直的且方向又与其轴线平行的有啮合零件的齿轮称作直齿轮。一对直齿轮只能用来连接平行轴。然而，平行轴也可以用其他类型的齿轮来连接，一个直齿轮可以和一个不同类型的齿轮相啮合。

第四节　轴系零部件
Section 4　Component of Shafting

滑动轴承　plain bearing/sliding-contact bearing
整体式滑动轴承　solid bearing
剖分式滑动轴承　split plain bearing
径向滑动轴承　plain sliding bearing
止推滑动轴承　plain thrust bearing
径向—止推滑动轴承　thrust-sliding bearing
液体动压滑动轴承　hydrodynamic bearing
液体静压滑动轴承　hydrostatic bearing
自润滑滑动轴承　self-lubricating bearing
自动调心滑动轴承　plain self-aligning bearing
轴瓦　liner
轴承衬　bearing liner
油槽　oil groove
油孔　oil hole
轴承径向载荷　bearing radial load
轴承轴向载荷　bearing axial load
轴承承载能力　bearing load carrying capacity
轴承压强　bearing mean specific load
滚动轴承　rolling bearing
单列轴承　single row bearing
双列轴承　double row bearing
角接触轴承　angular contact bearing
调心轴承　self-aligning bearing
推力轴承　thrust bearing
角接触推力轴承　angular contact bearing
单向推力轴承　single direction thrust bearing
双向推力轴承　double direction thrust bearing
球轴承　ball bearing
深沟球轴承　deep groove ball bearing
向心滚子轴承　radial roller bearing
圆柱滚子轴承　cylindrical roller bearing
圆锥滚子轴承　tapered roller bearing
滚针轴承　needle roller bearing
调心滚子轴承　self-aligning roller bearing
滚动体　rolling element
保持架　cage
轴承系列（滚动轴承）　bearing series (rolling bearing)
尺寸系列（滚动轴承）　dimension series (rolling bearing)
直径系列（滚动轴承）　diameter series (rolling bearing)
宽度系列（滚动轴承）　width series (rolling bearing)
轴承宽度　bearing width
轴承内径　bearing bore diameter

轴承外径	bearing outside diameter	轮胎式联轴器	coupling with rubber type element
静载荷	static load		
动载荷	dynamic load	弹性套柱销联轴器	pin coupling with elastic sleeves
当量载荷	equivalent load		
额定寿命	rating life	梅花形弹性联轴器	elastic pin coupling
基本额定寿命	basic rating life	离合器	clutch
径向载荷系数	radial load factor	操纵离合器	controlled clutch
轴向载荷系数	axial load factor	电磁离合器	electromagnetic clutch
联轴器	coupling	超越离合器	overrunning clutch
刚性联轴器	rigid coupling	滚柱离合器	roller clutch
套筒联轴器	sleeve coupling	安全离合器	safety clutch
凸缘联轴器	flange coupling	单向离合器	one-way clutch
夹壳联轴器	split coupling	摩擦式离合器	friction clutch
齿式联轴器	gear coupling	磁粉离合器	magnetic powder clutch
十字滑块联轴器	oldham coupling	制动器	brake
滑块联轴器	clao type coupling	牙嵌式制动器	jaw brake
链条联轴器	chain coupling	带式制动器	band brake
万向联轴器	universal joint	摩擦制动器	friction brake
牙嵌式联轴器	jaw and toothed coupling		

The concern of a machine designer with ball and roller bearings is fivefold as follows. (a) life in relation to load; (b) stiffness, i. e. defections under load; (c) friction; (d) wear; (e) noise. For moderate loads and speeds the correct selection of a standard bearing on the basis of load rating will usually secure satisfactory performance. The deflection of the bearing elements will become important where loads are high, although this is usually of less magnitude than that of the shafts or other components associated with the bearing. Where speeds are high special cooling arrangements become necessary which may increase frictional drag. Wear is primarily associated with the introduction of contaminants, and scaling arrangements must be chosen with regard to the hostility of the environment.

对于球轴承和滚子轴承,一个机器设计人员应该考虑下面五个方面。(a) 寿命与载荷的关系;(b) 刚度,也就是在载荷作用下的变形;(c) 摩擦;(d) 磨损;(e) 噪音。对于中等载荷和转速,根据额定负荷选择一个标准轴承,通常都可以保证其具有令人满意的工作性能。当载荷较大时,轴承零件的变形尽管通常小于轴和其他与轴承一起工作的零部件的变形,但将会变得重要起来。在转速高的场合需要有专门的冷却装置,而这可能会增大摩擦阻力。磨损主要是由于污染物的进入引起的,必须选用密封装置以防止周围环境的不良影响。

第五节 其他零部件
Section 5　Miscellaneous Component

弹簧	spring		cal helical spring
螺旋弹簧	helical spring	碟形弹簧	belleville spring
圆柱螺旋压缩弹簧	cylindrical helical compression spring	环形弹簧	ring spring
		板弹簧	leaf spring
圆柱螺旋拉伸弹簧	cylindrical helical tension spring	等刚度弹簧	constant rate (stiffness) spring
		变刚度弹簧	variable rate (stiffness) spring
圆柱螺旋扭转弹簧	cylindrical helical torsion spring	组合弹簧	combined spring
		自由高度(弹簧)	free height (spring)
不等节距圆柱螺旋弹簧	variable pitch cylindri-	工作高度(弹簧)	working height (spring)

工作载荷　specified load
极限载荷　ultimate load
总圈数　total number of coils
有效圈数　effective number of coil
极限高度　height under ultimate load

弹簧中径　mean diameter of coil
旋绕比（弹簧）　spring index (spring)
高径比（弹簧）　slender ratio (spring)
扭转角　torsion angle

Springs are mechanical members which are designed to give a relatively large amount of elastic deflection under the action of an externally applied load. Hooke's Law, which states that deflection is proportional to load, is the basis of behavior of springs. However, some springs are designed to produce a nonlinear relationship between load and deflection.

弹簧是一种能够在外载荷作用下，产生相当大的弹性变形的机械零件。变形与载荷成正比的胡克定律表明了弹簧的基本性能。然而，也有一些弹簧在其设计时所确定的载荷与变形之间的关系是非线性的。

第四章 材料力学
Chapter 4 Mechanics of Materials

第一节 绪论
Section 1 Introduction

材料力学	mechanics of Materials	各向同性	isotropy
构件	structure member	小变形	small deformations
强度	strength	内力	internal force
刚度	rigidity	外力	external force
稳定性	stability	二力构件	two force member
变形固体	solid deformation body	二力杆	two force bar
基本假设	fundamental assumption	极限强度	ultimate strength
连续性	continuity	刚体	rigid body
均匀性	homogeneity		

1 Strength — Capacity to resist fail of a component or an element.
强度——材料在外载荷作用下抵抗断裂或过量塑性变形的能力。

2 Rigidity — Capacity to resist deformation of a components or an element.
刚度——材料在外载荷作用下抵抗变形的能力。

3 Stability — Capacity to remain original state in the equilibrium of a component or an element.
稳定性——材料在外载荷作用下维持其原来平衡状态的能力。

4 Continuity — The material of a solid deformable body is continuously distributed over its volume so that there are not any cracks, defects or holes, etc..
连续性——物质密实地充满物体所在空间,无裂纹、缺陷或孔。

5 Homogeneity — The material of the solid deformable body is homogeneously distributed over its volume so that the smallest element cut from the body possesses the same specific mechanical properties as the body.
均匀性——材料在可变形固体的体积内均匀分配,因此从该物体上切下的最小部分拥有与该物体相同的机械特性。

6 Isotropy — The mechanical properties are the same in all directions at a point. Material with this property is called isotropy material. Material that the mechanical properties are different in all directions at a point is called anisotropy material.
各向同性——组成物体的材料沿各方向的力学性质完全相同。有这种性质的材料称为各向同性材料;沿各方向的力学性质不同的材料称为各向异性材料。

7 Small·deformations — The deformations for a solid deformable body caused by external forces are very small compared with the dimensions of the body. Thus when we study the equilibrium and motion of the deformable body, the deformation of the body may be neglected.
小变形——材料力学所研究的构件在载荷作用下的变形与原始尺寸相比甚小,故对构件进行受力分析时可忽略其变形。

第二节 轴向拉伸和压缩
Section 2 Axial Tension and Compression

轴向拉伸	axial tension	拉力	tensile force
轴向压缩	axial compression	内力	internal force

第四章 材料力学

截面法	method of section
轴力	axial force
轴力图	diagram of axial force
集中载荷	concentrate load
应力	stress
平均应力	average stress
全应力	whole stress
平面假设	hypothesis of plane section
拉应力	tensile stress
危险截面	critical section
危险点	critical point
圣维南原理	Saint-Venant principle
应力集中	stress concentration
强度设计准则	criterion of strength design
校核强度	check the strength
许可载荷	permissible load
斜截面	inclined section
失效准则	failure criteria
剪切	shearing
正应力	normal stress
剪应力	shearing stress
线应变	linear strain
剪应变	angular strain
剪力	shearing force
本构关系	constitutive relations
胡克定律	Hook law
弹性应变能	elastic strain energy
静定问题	statically determinate problem
静不定问题	statically indeterminate problem
平衡方程	equilibrium equation
变形协调方程	compatibility equation of deformation
物理方程	physical equation
装配应力	assemble stress
初始应力	initial stress
温度应力	temperature stress
力学性质	mechanical properties
静载荷	static load
低碳钢	low carbon steel
比例极限	proportional limit
弹性极限	elastic limit
屈服极限	yielding limit
塑性材料	ductile material
强度极限	strength limit
卸载定理	unloaded law
冷作硬化	cold hardening
颈缩	necking
延伸率	residual relative elongation
截面收缩率	permanent relative reduction of area
脆性	brittleness
塑性	ductility
名义屈服应力	nominal yield stress
铸铁	cast iron
剪切面	shearing plane
剪切破坏	failure due to shear
挤压破坏	breakage due to bearing
挤压面积	bearing area

1 Axial tension — Deformation of the rod is axial elongation and lateral shortening.
  轴向拉伸 —— 杆的变形是轴向伸长，横向缩短。

2 Axial compression — Deformation of the rod is axial shortening and lateral enlargement.
  轴向压缩 —— 杆的变形是轴向缩短，横向变粗。

3 Internal force — Internal force is the resultant of internal forces, which is acting mutually between two neighbor parts inside the body, caused by the external forces.
  内力 —— 指由外力作用所引起的、物体内相邻部分之间分布内力系的合成（附加内力）。

4 Axial force — Internal force of the rod in axial tension or compression, designated by N.
  轴力 —— 轴向拉压杆的内力，用 N 表示。

5 Hypothesis of plane section — Cross sections which were planes before deformation remain planes during and after deformation too. Deformations of elongate fibers are the same.
  平面假设 —— 原为平面的横截面在变形后仍为平面。纵向纤维变形相同。

6 Critical section — The section in which internal force is maximum and which dimension is smallest.
  危险截面 —— 内力最大的面，截面尺寸最小的面。

7 Critical point — The point stress is maximum.
  危险点 —— 应力点达到最大。

8 Saint-Venant principle — Distribution and magnitude of the stress in the section at a certain distance from the point at which the load is acted are not affected by the acting form of external loads.

圣维南原理——离开载荷作用处一定距离,应力分布与大小不受外载荷作用方式的影响。

9 The structure member that connects one member to another is called connecting member. Such as bolts, rivets, keys, etc.. Connecting member is small, but it plays passing loads role.

在构件连接处起连接作用的部件称为连接件。例如螺栓、铆钉、键等。连接件虽小,却起着传递载荷的作用。

第三节 扭转
Section 3 Torsion

轴 shaft	扭矩图 internal torque diagram
扭转 torsion	剪应力互等定理 theorem of complementary shearing stress
扭转角 angle of twist	
剪应变 shearing strain	轴力 axial force
扭矩 internal torque	

1 Shaft — In engineering the member which deformation is mainly torsion. Such as transmission shaft in machines, drill rod in oil-drilling rigs, etc..

轴——工程中以扭转为主要变形的构件。如机器中的传动轴、石油钻机中的钻杆等。

2 Torsion — Resultant of the external forces is a force couple and its acting plane is perpendicular to the axis of the shaft. Under this case deformation of the rod is torsion.

扭转——外力的合力为力偶,且力偶的作用面与直杆的轴线垂直,杆发生的变形为扭转变形。

3 Angle of twist — The angle of rotation of one section with respect to another.

扭转角——任意两截面绕轴线转动而发生的角位移。

4 Shearing strain — The change of a right angle between two straight lines.

剪应变——两条直线间直角的改变量。

5 Internal torque — The moment of internal forces acting in arbitrary section of the member in torsion. Designated by "T".

扭矩——构件受扭时,横截面上的内力偶矩,记作"T"。

6 Internal torque diagram — Sketch that expresses the law of change of the torque in each cross section along the axis.

扭矩图——表示沿杆件轴线各横截面上扭矩变化规律的图线。

7 Theorem of complementary shearing stresses — It indicates shearing stresses always exist on mutually perpendicular plane and occur in equal and opposite pairs and point, perpendicularly, either toward or away from the line of intersection of the planes.

剪应力互等定理——该定理表明:在单元体相互垂直的两个平面上,剪应力必然成对出现,且数值相等,两者都垂直于两平面的交线,其方向则共同指向或共同背离该交线。

第四节 弯曲内力
Section 4 Internal Forces in Bending

弯曲 bending	简支梁 simply supported beam
梁 beam	悬臂梁 cantilever beam
平面弯曲 planar bending	外伸梁 overhanging beam
固定铰支座 fixed hinged support	曲杆 curved rod
可动铰支座 movable hinged support	静定梁 statically determinate beam
固定端 rigidly fixed end	超静定梁 statically indeterminate beam

弯曲内力　internal force in bending	弯矩图　bending moment diagram
弯矩　bending moment	分布荷载　distributed load
内力方程　internal force equation	荷载集度　density of the distributed load
剪力图　shearing force diagram	

1 Bending: The action of the force and external couple vector perpendicular to the axis of the rod makes the axis of the rod change into curve from original straight line, this deformation is called bending.

　　弯曲——杆受垂直于轴线的外力或外力偶矩矢的作用时，轴线变成了曲线，这种变形称为弯曲。

2 Beam: The member in which deformation is mainly bending is generally called beam.

　　梁——以弯曲变形为主的构件通常称为梁。

3 Planar bending: After deformation the curved axis of the beam is still in the same plane with the external forces.

　　平面弯曲——杆发生弯曲变形后，轴线仍然和外力在同一平面内。

4 Theorem of superposition: Internal forces in the structure due to simultaneous action of many forces are equal to algebraic sum of the internal forces due to separate action of each force.

　　叠加原理——多个载荷同时作用于结构而引起的内力等于每个载荷单独作用于结构而引起的内力的代数和。

第五节　弯曲应力
Section 5　Stress in Bending

中性层　neutral layer	等强度梁　equal strength beam
中性轴　neutral axis	主惯性轴　principal axis of inertia
抗弯截面模量　section modulus of bending	弯曲中心　bending center
曲率半径　radius of curvature	正应力分布　distribution of normal stress
圆截面　circular section	静力关系　static relation
薄壁圆环　thin-walled cirque	物理关系　physical relation
槽钢　channel steel	弯心　bending center
变截面梁　beam of change section	纯弯曲　pure bending

1 Pure Bending: Deformation of some portion of the beam in which there are only bending moment and no shearing stress is called pure bending.

　　纯弯曲——某段梁的内力只有弯矩没有剪力时，该段梁的变形称为纯弯曲。

2 Neutral layer: A layer at a certain height inside the beam in which the longitudinal fibers are neither to be elongated nor to be shortened, neither subject to tension nor compression. This layer is called neutral layer.

　　中性层——梁内一层纤维既不伸长也不缩短，因而纤维不受拉应力和压应力，此层纤维称中性层。

3 Neutral axis: The intersection of the neutral layer with any cross section.

　　中性轴——中性层与横截面的交线。

第六节　弯曲变形
Section 6　Deformation in Bending

挠度　deflection	位移边界条件　boundary conditions of the displacement
转角　angle of rotation	
挠曲线　deflection curve	支点位移条件　displacement conditions at the supports
挠曲线近似微分方程　approximate differential equation of deflection curve	连续条件　continuity condition
弹性曲线　elastic curve	光滑条件　smooth condition

积分常数　integral constant
载荷叠加　superposition of loads
许可转角　permissible angle of rotation
弯曲应变能　strain energy in bending

1 Study range — Calculation of the displacement of the straight beam with equal section in symmetric bending.
研究范围——等直梁在对称弯曲时位移的计算。

2 Study object — ①Do rigidity check for the beam;②Solve problems about statically indeterminate beams (complementary equations are supplied by the conditions of deformation of the beam).
研究目的——① 对梁作刚度校核；② 解决静定梁问题（变形几何条件提供补充方程）。

3 Deflection — The displacement of the centroid of a section in a direction perpendicular to the axis of the beam. It is designated by the letter v. It is positive if its direction is the same as f, or negative.
挠度——横截面形心沿垂直于轴线方向的线位移。用 v 表示。与 f 同向为正，反之为负。

4 Angle of rotation — The angle by which cross section turns with respect to its original position about the neutral axis. It is designated by the letter θ. It is positive if the angle of rotation rotates in the clockwise direction, or negative.
转角——横截面绕其中性轴转动的角度。用 θ 表示，顺时针转动为正，反之为负。

5 Approximate differential equation of the deflection curve — The curve which the axis of the beam was transformed into after deformation is called deflection curve. Its equation is $v = f(x)$.
挠曲线近似微分方程——变形后，轴线变为曲线，该曲线称为挠曲线。其方程为：$v = f(x)$。

第七节　应力状态和强度理论
Section 7　Analysis of Stressed State and Strength Theories

一点的应力状态　stressed state at a point
原始单元体　original element
主单元体　principal element
主面　principal plane
主应力　principal stress
三向应力状态　three dimensional state of stress
二向应力状态　plane state of stress
单向应力状态　unidirectional state of stress
极值应力　extreme values for the stress
应力圆　stress circle
主应力迹线　principal stress trajectories
空间应力状态　spatial stressed state
复杂应力状态　complex stress state
体积应变　volumetric strain
应力分量　stress component
内压力　inside compressive force
应变能密度　strain-energy density
形状改变能密度　strain-energy density corresponding to the distortion
强度理论　theories of strength
最大拉应力（第一强度）理论　theory of maximum tensile stress (first strength)
破坏判据　criterion of rupture
强度准则　strength condition
最大伸长线应变（第二强度）理论　theory of maximum tensile strain (second strength)
最大剪应力（第三强度）理论　theory of maximum shearing stress (third strength)
形状改变比能（第四强度）理论　theory of maximum torsional (shearing) strain energy (fourth strength)
极限应力圆　circle of limit stress
极限曲线　limit curve
极限应力圆的包络线　envelope of circles of limit stress
莫尔强度理论　Mohr's strength theory
相当应力　equivalent stress

1 Stressed state at a point — There are countless section through a point. Set of stresses in all section is called stressed state at this point.
一点的应力状态——过一点有无数的截面，这一点的各个截面上应力情况的集合，称为这点的应力状态。

2 Principal element — The element in which the shearing stresses in side planes are all zero.
主单元体——各侧面上剪应力均为零的单元体。

3. Principal plane — The plane in which the shearing stresses are totally absent.
主面 —— 剪应力为零的截面。
4. Principal stress — Normal stresses act in the principle plane.
主应力 —— 主面上的正应力。
5. Three Dimensional State of Stress — State of stress in which all the three principal stresses are not equal to zero.
三向应力状态 —— 三个主应力都不为零的应力状态。
6. Plane State of Stress — State of stress in which one principal stresses is equal to zero.
二向应力状态 —— 一个主应力为零的应力状态。
7. Unidirectional State of Stress — State of stress in which one principal stresses is not equal to zero.
单向应力状态 —— 一个主应力不为零的应力状态。
8. Theories of strength — Some assumptions about the cause of the failure of materials.
强度理论 —— 关于"构件发生强度失效起因"的假说。
9. Theory of maximum tensile stress (first strength) — This theory thinks the main cause of rupture is the maximum tensile stress. The member rupture as the maximum tensile stress reaches the strength limit in axial tension.
最大拉应力(第一强度)理论 —— 认为构件的断裂是由最大拉应力引起的。当最大拉应力达到单向拉伸的强度极限时,构件就断了。
10. Theory of maximum tensile strain (second strength) — This theory thinks the main cause of rupture is the maximum tensile strain. The member rupture as the maximum tensile strain reaches the limit strain in axial tension.
最大伸长线应变(第二强度)理论 —— 认为构件的断裂是由最大拉应力引起的。当最大伸长线应变达到单向拉伸试验下的极限应变时,构件就断了。
11. Theory of maximum shearing stress (third strength) — This theory thinks the main cause of rupture is the maximum shearing stress. The member rupture as the maximum shearing stress reaches the limit stress in axial tension.
最大剪应力(第三强度)理论 —— 认为构件的屈服是由最大剪应力引起的。当最大剪应力达到单向拉伸试验的极限剪应力时,构件就破坏了。
12. Theory of maximum torsional (shearing) strain energy (fourth strength) — This theory thinks the main cause of yield is the maximum torsional strain energy. The member rupture as the maximum torsional strain energy reaches the torsional strain energy of yield in axial tension.
形状改变比能(第四强度)理论 —— 认为构件的屈服是由形状改变比能引起的。当形状改变比能达到单向拉伸试验屈服时形状改变比能时,构件就破坏了。
13. Mohr's strength theory — The stress circle of arbitrary point contacted with the limit curve material will be yield or cut off.
莫尔强度理论 —— 任意一点的应力圆若与极限曲线相接触,则材料即将屈服或剪断。
14. Mohr thought the maximum shearing stress is main cause of failure of material but friction force in the sliding section is not neglected (Law of Mohr friction). Combined the factors of the maximum shearing stress and the maximum normal stress Mohr obtained his strength theory.
莫尔认为:最大剪应力是使物体破坏的主要因素,但滑移面上的摩擦力也不可忽略(莫尔摩擦定律)。综合最大剪应力及最大正应力的因素,莫尔得出了他自己的强度理论。

第八节 组合变形
Section 8　Composite Deformation

组合变形　composite deformation	斜弯曲　skew bending
中性轴方程　equation of the neutral axis	截面核心　kernel of section

1 The members of structures give rise to two or more types of simple deformation due to complex external loads simultaneously. If stresses corresponding to each simple deformation can not be neglected when their magnitudes belong to the same magnitude, this kind of deformation is called composite deformation.

在复杂外载荷作用下,构件的变形会包含几种简单变形,当几种变形所对应的应力属同一量级时,不能忽略之,这类构件的变形称为组合变形。

2 Skew bending — After bending deformation of the rod the curve of deflection and the external forces (transversal forces) are not in the same plane.

斜弯曲——杆件产生弯曲变形,但弯曲后,挠曲线与外力(横向力)不共面。

3 Bending and tension or compression — Deformation of the rod due to simultaneous action of transversal and axial forces.

拉(压)弯组合变形——杆件同时受横向力和轴向力的作用而产生的变形。

第九节 压杆稳定性
Section 9 Stabilization of Compressive Columns

构件的承载能力 load-carrying capacity of structure members
不稳定平衡 instable equilibrium
稳定平衡 stable equilibrium
压杆失稳 loss of stability of the column
临界压力 critical pressure
理想压杆 ideal compressive column
长度系数 length coefficient
约束系数 constraint coefficient
临界应力 critical stress
柔度 flexibility
大柔度杆 large flexibility column
中小柔度杆 columns with middle and small flexibility
临界应力总图 total figure of critical stress
抛物线形经验公式 empirical formula of a parabolic variation
压杆的稳定容许应力 permissible stress for the columns in stabilization
压杆的稳定条件 stability condition of the columns

1 Ideal compressive columns — material is absolutely ideal; the axis is absolutely straight; compressive force is along the axis of the column.

理想压杆——材料绝对理想;轴线绝对直;压力绝对沿轴线作用。

2 Critical stress — Mean stress in the cross section of the column in critical state.

临界应力——压杆处于临界状态时横截面上的平均应力。

3 Some structure members in engineering have enough strength and rigidity but they are not sure to work safely and reliably.

工程中有些构件具有足够的强度、刚度,却不一定能安全可靠地工作。

第十节 平面图形的几何性质
Section 10 Geometry Nature of Plan Figure

面积(对轴)矩 area (to axle) square
组合图形 combination graph
惯性矩 inertia couple
极惯性矩 polar moment of inertia
惯性积 inertia accumulating
转轴定理 pivot theorem
平行移轴定理 parallel moves the axis theorem
主惯性轴 main inertia axis
主惯性矩 main inertia couple
形心主轴 main shaft of heart of shape
形心主惯性矩 main inertia couple of heart of shape

1 Area (to axle) square — is the product of the area and the distance to the axis.

面积(对轴)矩——是面积与它到轴的距离之积。

2 Inertia couple is the product of the area and the square of the distance to the axis.
惯性矩是面积与它到轴的距离的平方之积。
3 Polar moment of inertia is the second moment of area of the pole.
极惯性矩是面积对极点的二次矩。
4 Inertia accumulating is the product of the area and the distance between two axis.
惯性积是面积与其两轴距离之积。

第五章　理论力学
Chapter 5　Theoretical Mechanics

第一节　绪论
Section 1　Introduction

静力学　statics
运动学　kinematics

动力学　dynamics
理论力学　theoretical mechanics

1 Theoretical Mechanics — The science of the general laws of mechanical motions of material bodies.
　理论力学 —— 研究物体机械运动一般规律的一门学科。

2 Mechanical motion — Any change in the relative positions of material bodies in space which occurs in the course of time.
　机械运动 —— 物体在空间的位置随时间的变化。

3 Statics — Studies the laws of equilibria of material bodies subjected to the action of forces, the general character of forces and the methods to simplify force systems.
　静力学 —— 研究物体在力系作用下的平衡规律，同时也研究力的一般性质和力系的简化方法等。

4 Kinematics — Deals with the general geometrical description of the motions of bodies, does not study the reasons of their motions.
　运动学 —— 研究物体运动的几何性质，而不研究引起物体运动的原因。

5 Dynamics — Studies the laws of motion of material bodies under the action of forces.
　动力学 —— 研究受力物体的运动变化与作用力之间的关系。

第二节　静力学的基本原理
Section 2　Fundamental Principles of Statics

力　force
力系　force system
平衡力系　balanced force system
刚体　rigid body
平衡状态　state of equilibrium
公理　axion
二力平衡公理　axion of two forces equilibrium
二力体　double forces equivalent body
加减平衡力系公理　add-subtract balance force system
力的可传性　transmissibility of force
力的平行四边形法则　law of parallelogram
三力平衡汇交定理　theorem of three equivalent forces intersecting at one point
作用力与反作用力定律　law of action and reaction force

自由体　free body
非自由体　constrained body
约束　constrain
约束反力　reaction force of constraint
光滑接触面约束　constrain of smooth interface
光滑圆柱铰链约束　constrain of smooth cylindrical gemel
圆柱铰链　cylindrical gemel
固定铰支座　fixed hinged support
滑槽　runner
活动铰支座　movable hinged support
辊轴支座　bearing of roll shaft
主动力　active force
被动力　passive force
受力图　force diagram
刚化原理　principle of rigidization

1 Force — The mechanical interaction of bodies is caused by forces.
　力 —— 力是物体间的相互机械作用。

2 Force system — A group of force acts on a body.
　力系 —— 是指作用在物体上的一群力。

第五章 理论力学

3 Balanced force system — If a body subjected to a force system is in equilibrium state, we call the system to be a balanced force system.
平衡力系 —— 物体在力系作用下处于平衡状态，这个力系称为平衡力系。

4 Rigid body — A body which does not change its shape and dimension under applied forces is called a rigid body.
刚体 —— 就是在力的作用下，大小和形状都不变的物体。

5 Equilibrium state — A body is called to be in the state of equilibrium if it does not move or if it moves with a uniform velocity.
平衡 —— 是指物体相对于惯性参考系保持静止或作匀速直线运动的状态。

6 Principle — Conclusions obtained from longtime practice and many experiments. They are repeatedly proved by practice and now are recognized without prove.
公理 —— 是人类经过长期实践和经验而得到的结论，它被反复的实践所验证，是无须证明而为人们所公认的结论。

7 The law of parallelogram — Two forces applied at one point of a body have as their resultant a force applied at the same point and represented by the diagonal of a parallelogram constructed with two given forces as its sides.
力的平行四边形法则 —— 作用于物体上同一点的两个力可合成一个合力，此合力也作用于该点，合力的大小和方向由以原两力矢为邻边所构成的平行四边形的对角线来表示。

8 Three equivalent forces intersect at one point — If a free rigid body remains in equilibrium under the action of three nonparallel coplanar forces, the lines of action of those forces intersect at one point. Moreover, three forces are coplanar (under special circumstances, forces intersect at infinite and become a parallel force system).
三力平衡汇交定理 —— 刚体受三力作用而平衡，若其中两力作用线汇交于一点，则另一力的作用线必汇交于同一点，且三力的作用线共面（必共面，在特殊情况下，力在无穷远处汇交并且成为平行力系）。

9 The law of action force and reaction force — To any action of one material body on another there is always an equal and oppositely directed reaction. They are equal in magnitude, opposite in direction, collinear and exist together, but act on different body.
作用力和反作用力定律 —— 一个物体对另一物体的作用总会呈现同方向或相反方向的反应。它们等值、反向、共线、异体，且同时存在。

第三节 平面特殊力系
Section 3 Special Cases of Force Systems in a Plane

平面力系 coplanar force system
平面一般力系 general case of force system in plane
平面平行力系 coplanar system of parallel forces
力偶 force couple
平面汇交力系 coplanar system of concurrent forces
平面特殊力系 special cases of force system in plane
解析法 analytical method

合力投影定理 law of projection of a resultant force
合力矩定理 law of the moment of a resultant force
平面力偶 coplanar for couple
平面力偶等效定理 equivalent law of coplanar fore couple
平面力偶系 coplanar system of force couple
力对点的矩 moment of force relative to a point

1 Special case of force system in a plane — Coplanar system of concurrent forces, coplanar system of force couples and coplanar system of parallel forces.
平面特殊力系 —— 指的是平面汇交力系、平面力偶系和平面平行力系。

2 The law for the moment of a resultant force — The moment of the resultant of a coplanar system of concurrent forces about any center is equal to the algebraic sum of the moments of the component forces about that center.
合力矩定理 —— 平面汇交力系的合力对平面内任一点的矩，等于所有各分力对同一点的矩的代数和。

3 Force couple — A system of two parallel forces of the same magnitude and opposite direction acting on a rigid body.
力偶 —— 两力大小相等，方向相反作用于刚体的两个平行力系叫力偶。

4 Coplanar system of force couples — Many force couples acting on a rigid body lie in the same plane.
平面力偶系 —— 在同一个平面内，同时作用于一个刚体之上的各力之合。

第四节　一般力系
Section 4　General Coplanar Force System

力线平移定理　theorem of translation of a force	三矩式平衡方程　balanced equation with three moments
主矢　principle vector	
主矩　principle moment	充要条件　necessary and sufficient condition
简化中心　center of reduction	物体系统的平衡问题　equilibrium of a body system
车刀　turning tool	
作用线　action line	桁架　truss
一矩式平衡方程　balanced equations with one moment	节点法　node method
二矩式平衡方程　balanced equation with two	连续梁　continuous beam

1 Theorem of translation of a force — A force acting on a rigid body can be moved parallel to its active line to any point of the body, if we add a couple with a moment equal to the moment of the force about the point to which it is translated.
力的平移定理 —— 作用于刚体的力可平移到该刚体的任何一点，但必须同时附加一个力偶，这个力偶的矩等于原来的力到新移动点之矩。

2 Truss — A rigid structure system composed of straight elements, connected at their ends by pins, without deformation under the action of forces.
桁架 —— 由杆组成，用铰连接，受力不变形的刚性结构系统。

第五节　质点的运动
Section 5　Kinematics of a Particle

运动方程　equation of motion	柱坐标　column coordinate
运动轨迹　motion trajectory	切向加速度　tangential acceleration
质点的速度　velocity of a particle	法向加速度　normal acceleration
加速度　acceleration	加速运动　accelerated motion
弧坐标　curvilinear coordinate	匀速曲线运动　curvilinear motion of uniform velocity
自然轴系　natural coordinate system	
极坐标　polar coordinate	减速曲线运动　decelerated curvilinear motion

1 Tangential acceleration — It represents the rate of change.
切向加速度 —— 表示速度大小的变化。

2 Normal acceleration — It represents the rate of change of speed direction of the particle.
法向加速度 —— 表示速度方向的变化。

第六节 刚体的基本运动
Section 6　Basic Motion of a Rigid Body

直线平动　rectilinear translation	eration
曲线平动　curvilinear translation	线速度　linear velocity
定轴转动　rotation about a fixed axis	齿轮传动　transmission of gear
刚体平动　translational motion of a rigid body	内啮合　inner meshing
转动方程　equation of rotation	齿轮传动比　gear ratio
转角　angle of rotation	外啮合　external meshing
逆时针　anticlockwise	皮带轮系传动　transmission system consisting of blet and puller
顺时针　clockwise	
角速度　angular velocity	链轮系　system of sprocket wheel
角加速度　angular acceleration	右手定则　right-hand rule
匀速转动　rotation with uniform velocity	全加速度　substantive acceleration
匀变度转动　rotation with uniform angular accel-	

1 The shape and dimensions of a rigid body should be considered investigating its motion. However, due to its unchanged shape, it is not necessary to determine its position by finding the positions of all its points. Instead, its position is determined completely by describing the position of a line or a plane in it.

由于研究对象是刚体,所以运动中要考虑其本身形状和尺寸大小,又由于刚体是几何形状不变的,所以研究它在空间中的位置就不必一个点一个点地确定,只要根据刚体的各种运动形式,确定刚体内某一个有代表性的直线或平面的位置即可。

2 Features of translational motion: All the particles of a rigid body with translational motion have the same motion, i.e. their trajectories, velocities and accelerations are identical. The translation of a rigid body can be simplified to the motion of a particle in it.

刚体平动的特点:平动刚体在任一瞬时各点的运动轨迹形状、速度、加速度都一样。刚体平动可以简化为质点的运动。

第七节 质点运动的合成
Section 7　Compositive Motion of a Particle

坐标系　coordinate system	绝对加速度　absolute acceleration
静坐标系　static coordinate system	相对加速度　relative acceleration
动坐标系　moving coordinate system	桥式吊车　overhead traveling crane
动点　moving point	牵连加速度　convected acceleration
绝对运动　absolute motion	圆盘凸轮机构　disk cam mechanism
相对运动　relative motion	速度合成定理　theorem of velocity composition
牵连运动　convected motion	加速度合成定理　theorem of acceleration composition
牵连点　convected point	

1 Static coordinate system — A coordinate system fixed to the earth ground is called Static coordinate system (SCS).

静坐标系——把固结于地面上的坐标系称为静坐标系,简称静系。

2 Moving coordinate system — A coordinate system fixed to a moving object relative to the earth ground is called moving coordinate system (MCS). For example, a running car.

动坐标系——把固结于相对于地面运动物体上的坐标系,称为动坐标系,简称动系。例如在行驶的汽车。

第八节 刚体的平面运动
Section 8 Plane Motion of a Rigid Body

平面运动方程　equations of plane motion	瞬时平动　instantaneous translation
曲柄连杆机构　crank-rod mechanism	速度瞬心　instantaneous velocity center
基点法　pole-based method	行星齿轮机构　mechanism of planet gears
速度投影法　velocity projection method	导槽滑轮机构　slotted bar-sliding block mechanism
瞬时速度中心　instantaneous velocity center	

1 In engineering, we often encounter the plane motion of a rigid body, which is more complicated. Investigation into this kind of motion is conducted on the basis of translation and rotation of a rigid body using composition and decomposition of motions. Decomposing a plane motion into translation and rotation, and employing the theory of composition of motions, the formula for finding the velocity and acceleration of point in the rigid body can be derived.

刚体的平面运动是工程上常见的一种运动,这是一种较为复杂的运动。对它的研究可以在研究刚体的平动和定轴转动的基础上,通过运动合成和分解的方法,将平面运动分解为上述两种基本运动。然后应用合成运动的理论,推导出平面运动刚体上一点的速度和加速度的计算公式。

2 The distance between any point in a rigid body and a fixed plane always keeps unchanged during its motion. In other words, any point in the rigid body moves in a plane parallel to the fixed plane. The motion described above is called plane motion of a rigid body.

在运动过程中,刚体上任一点到某一固定平面的距离始终保持不变。也就是说,刚体上任一点都在与该固定平面平行的某一平面内运动。具有这种特点的运动称为刚体的平面运动。

3 A plane motion of a rigid body can be simplified to a motion of a plane figure in the plane itself. As indicated in the figure, when studying a plane motion of a rigid body, we do not need to consider its geometrical shape and size are not needed, and instead, considering the motion of a plane figure is enough to determine the velocity and acceleration of any point in the rigid body.

刚体的平面运动可以简化为平面图形在其自身平面内的运动。即在研究平面运动时,不需考虑刚体的形状和尺寸,只需研究平面图形的运动,确定平面图形上各点的速度和加速度。

第九节 刚体的一般运动
Section 9 General Motion of a Rigid Body

刚体绕平行轴转动的合成　composition of rotation around two parallel axis	绝对角速度　absolute angular velocity
	瞬时轴　instantaneous axis
速度中心　velocity center	牵连角速度　embroiling angular velocity

Composition problems of rotation around parallel axis is very common in mechanical engineering. For instance, the planet gear in a planet gear mechanism has plane motion. In previous study, we dealt with a plane motion by decomposing it into translation and rotation. However, it is more convenient to view the plane motion of a planet gear as a composition of rotation.

绕平行轴转动的合成问题在机械中经常遇到。例如,在行星齿轮机构中行星齿轮可作平面运动。在前面的研究中,我们把平面运动分解成平动和转动。然而,在分析行星轮系的传动问题时,将行星轮的平面运动看成为行星齿轮与转动的合成运动则比较方便。

第十节 质点运动微分方程
Section 10 Differential Equations of the Motion of a Particle

矢量形式　vector form	直角坐标形式　forms of rectangular coordinates
自然形式　natural form	

1 The basic equations of dynamics represented by equations in differential forms is called the differential equations of motion of a particle.

将动力学基本方程表示为微分形式的方程,称为质点运动微分方程。

2 Besides these three forms, the differential equations of motion of a particle may be represented in the forms of polar and other coordinates. Applying the equations we can solve two types of problems of particle dynamics.

质点运动微分方程除以上三种基本形式外,还可有极坐标形式、柱坐标形式等。应用质点运动微分方程,可以求解质点动力学的两类问题。

第十一节 动量定理
Section 11　Theorem of Momentum

动力学普遍定理　general theorems of dynamics
动量定理　theorem of momentum
动能定理　theorem of kinetic energy
质点系的质心　center of mass of a system of particles
动量　momentum
质点的动量　momentum of a particle
质点系的动量　momentum of a system of particles
刚体系统的动量　momentum of a system of rigid bodies
冲量　impulse
常矢量　constant vector
变矢量　variable vector
元冲量　elementary impulse
合理的冲量　impulse of a resultant force
质点的动量定理　theorem of momentum for one particle
质点的动量守恒　law of conservation of the momentum of a particle
质点系的动量定理　theorem of momentum of a system of particles
质点系的动量守恒定律　conservation law of th linear momentum of a system of particles
刚体系统　system of rigid bodies
质心运动定理　theorem of motion of the center of mass
质心运动守恒定律　conservation law of the momentum of the center of mass

1 In a homogeneous gravitational field the centers of mass and the center gravity coincide. Using any method of determination of the center of gravity determines the position of the center of mass of a system. But they are not identical. The concept of the center of mass has more mechanical meaning than that of the center of gravity.

在均匀重力场中,质点系的质心与重心的位置重合。可采用静力学中确定重心的各种方法来确定质心的位置。但是,质心与重心是两个不同的概念,质心比重心具有更加广泛的力学意义。

2 Momentum of a particle: The product (mv) of the mass of a particle and its velocity is called the momentum of a particle. It is a time-dependent vector with the same direction as the velocity, the unit of which is kg·m/s.

质点的动量:质点的质量与速度的乘积(mv)称为质点的动量,它是瞬时矢量,方向与速度相同,单位是 kg·m/s。

3 Momentum is a physical quantity measuring the intensity of the mechanical motion of a material body. For example, the velocity of a bullet is big but its mass is small. In the case of a boat it is just opposite.

动量是度量物体机械运动强弱程度的一个物理量。例如,枪弹的速度大,但质量小;船的速度小,但质量大。

4 Impulse: The product of a force and the action time of the force is called impulse. Impulse is used to characterize the accumulated effect on a body of a force acting during a certain time interval. For instance, when we push a cart, a larger force during a shorter time interval may obtain the same general effect as a smaller force during a longer time interval.

冲量：力与其作用时间的乘积称为冲量，冲量表示力在其作用时间内对物体作用的累积效应的度量。例如，推动车子时，较大的力作用较短的时间，与较小的力作用较长的时间，可得到同样的总效应。

第十二节　动量矩定理
Section 12　Moment of Momentum Theorem

质点的动量矩　moment of momentum of a particle

质点系的动量矩　moment of momentum of a system of particles

平动刚体　rigid body in translation motion

定轴转动刚体　rigid body in fixed-axis rotation

平面运动刚体　rigid body in plane motion

质点的动量矩定理　moment of momentum theorem of a particle

质点系的动量矩定理　moment of momentum theorem of a system of particles

刚体定轴转动微分方程　differential equation for the rotation of rigid body around a fixed-axis

回转半径　radius of gyration

平行轴定理　parallel axis theorem

质点系相对质心的动量矩定理　moment of momentum theorem of a system of particles with respect to its center of mss

刚体平面运动微分方程　differential equations of plane motion of a rigid body

1 The moment of momentum of a rigid body in translational motion with respect to a fixed center (axis) is equal to the moment of momentum of the center of mass of the rigid body with respect to that center (axis).

平动刚体对固定点（轴）的动量矩等于刚体质心对该点（轴）的动量矩。

2 The moment of momentum of a rigid body in fixed-axis rotation with respect to the axis of rotation is equal to the product of the moment of inertia of the rigid body with respect to that axis and its angular velocity.

定轴转动刚体对转轴的动量矩等于刚体对该轴转动惯量与角速度的乘积。

3 The moment of momentum of a rigid body in plane motion with respect to an axis perpendicular to the symmetrical plane of mass is equal to the sum of the moment of momentum of the center of mass of the rigid body in translational motion together with the center of mass about that axis and the moment of momentum of the rigid body when rotating around the center of mass with respect to a parallel axis through the center of mass of the body.

平面运动刚体对垂直于质量对称平面的固定轴的动量矩，等于刚体随同质心做平动时质心的动量对该轴的动量矩与绕质心轴做转动时的动量矩之和。

4 The moment of momentum theorem of a system of particles with respect to any fixed axis: The time-derivative of the total moment of momentum of a system with respect to any fixed axis is equal to the algebraic sum of the moments of all the external forces acting on the system with respect to the same axis (the principal moment of external force system with respect to the same axis). The theorem states that only external forces can change the moment of momentum of the system, internal forces ca not do this.

质点系对固定轴的动量矩定理：即质点系对任一固定轴的总动量矩对时间的导数，等于作用在质点系上的所有外力对同一固定轴之矩的代数和（外力系对同一轴的主矩）。该定理说明内力不会改变质点系的动量矩，只有外力才能改变质点系的动量矩。

5 The parallel axis theorem: The moment of inertia of a rigid body with respect to any axis is equal to the moment of inertia of the body with respect to a parallel axis through the center of mass of the body plus the product of the mass of the body with the square of the distance between the two axes.

平行移轴定理：刚体对某轴的转动惯量等于刚体对通过质心且与该轴平行的轴的转动惯量，加上刚体的质量与两轴间距离的平方之乘积。

第五章 理论力学

第十三节 动能定理
Section 13　Theorem of Kinetic Energy

力的功　work of a force
常力的功　work done by a constant force
变力的功　work done by a variable force
元功　elementary work
合力的功　work of a resultant force
重力的功　work of gravity
力偶的功　work of a force couple
弹性力的功　work of an elastic force
万有引力的功　work of universal gravitation
摩擦力的功　work of friction
滑动摩擦力的功　work of kinetic friction
质点系内力的功　work of internal force of a system of particles
向心轴承　centripetal bearing
理想约束反力的功　work of reaction forces of ideal constraints
动能　kinetic energy
质点的动能　kinetic energy of a particle
质点系的动能　kinetic energy of a system of particles
动能定理　theorem of kinetic energy
质点的动能定理　theorem of kinetic energy of a particle
质点系的动能定理　theorem of kinetic energy of a system of particles
功率方程　power equation
势力场　field of conservative force
势能　potential energy
机械能守恒定律　law of conservation of mechanical energy
重力场　field of gravity
弹性力场　field of elastic force
万有引力功　field of universal gravitation
有势力的功　works of conservative force
动力学普遍定律　general theorems of dynamics

1 The kinetic energy of a material body is the energy due to the motion, the kinetic energy is a measure of the mechanical motion.
物体的动能是由于物体运动而具有的能量,是机械运动的一种度量。

2 Power: Power is the work done by a force in a unit of time. It is an important index that measures the working capacity of a machine. Power is a scalar quantity and has a instantaneous characteristic.
功率:力在单位时间内所做的功。它是衡量机器工作能力的一个重要指标。功率是代数量,并有瞬时性。

3 Force field: If in any region of space a particle experiences a force of a certain magnitude and direction depending on position, the region of space is called to be a force field.
力场:若质点在某空间内的任何位置都受到一个大小和方向完全由所在位置确定的力的作用,则此空间称为力场。

4 Field of conservative force: If in a force field the work done by the force acting on a moving particle depends only on the initial and the final position of the particle and does not depend on path the force field is called to be a field of conservative force.
势力场:在力场中,如果作用于某运动质点的场力做功只取决于质点的始末位置,与运动路径无关,这种力场称为势力场。

第十四节 达朗伯原理
Section 14　D'Alembert's Principle

惯性力　inertial force
质点的达朗伯原理　D'Alembert's principle for a particle
主动力系　positive force system
惯性力系　inertial force system
质点系的达朗伯原理　D'Alembert's principle for a system of particles
约束反力系　system of reaction forces of constraint
刚体轴承动反力　dynamical reaction of the bearing of a rigid body

1 An important theorem of dynamics, D'Alembert's principle, will be introduced in this section. By this principle dynamical problems can be transformed formally into those of statics. Then they can be solved by the theorem of equilibrium. Therefore this method to solve the dynamical problems of dynamics is called the dynamic-static method.
本节介绍动力学的一个重要原理——达朗伯原理。应用这一原理，就将动力学问题从形式上转化为静力学问题，从而通过平衡定理来求解。因而这种解答动力学问题的方法被称为动静法。

2 The inertial force is not the real force acting on the particle, it is the resultant force of reaction of the particle to the object applying the force.
惯性力不是作用在质点上的真实力，它是质点对施力体反作用力的合力。

3 No matter what motion the rigid body is doing, the principle vector of the inertial force system is equal to the product of the weight of the rigid body and the acceleration of the center of mass, the direction is opposite to the direction of the acceleration of the center of mass.
无论刚体做什么运动，惯性力系主矢都等于刚体质量与质心加速度的乘积，方向与质心加速度方向相反。

第六章 机械制造技术基础及装备设计
Chapter 6　Fundamentals of Manufacturing Technology and Manufacturing Equipment Design

第一节　机械加工方法
Section 1　Machining Method

加工方法　machining operation
主运动　main motion
进给运动　feed motion
★ 车削　turning
★ 铣削　milling
顺铣　down milling
逆铣　up milling
★ 刨削　planing, planing and shaping
插削　slotting
拉削　broaching
铲削　relieving
★ 磨削　grinding
成形加工　forming
仿形加工　copying

★ 铰削　reaming
★ 镗削　boring
拉孔　hole broaching
螺纹加工　thread machining
齿轮加工　gear cutting
成形法　forming method
展成法　generating method
插齿　gear shaping
滚齿　gear hobbing
特种加工工艺　non-traditional machining
电火花加工　electro-discharge machining (EDM)
电解加工　electrochemical machining
激光加工　laser machining
超声波加工　ultrasonic machining

1 Machine tools are generally power-driven metal-cutting or forming machines used to shape metals by: a. the removal of chips, b. pressing, drawing, or shearing, c. controlled electrical machining processes.
机床是动力驱动的通过切除切屑的金属切削加工或通过挤压、拉伸或剪切的金属成形加工或电加工过程。

2 Of all the machining processes performed, drilling makes up about 25%. Consequently, drilling is a very important process.
在所有的机械加工方法中,钻削占 25%。因此,钻削是很重要的加工方法。

3 Milling operations can be classified into two broad categories called peripheral milling and face milling.
铣削加工可以分为圆周铣和平面铣两大类。

第二节　金属切削原理与刀具
Section 2　Metal Cutting Principle and Cutting Tools

★ 刀具　tool
★ 工件　workpiece
切削速度　cutting speed
进给量　feed
背吃刀量(切削深度)　depth of cut
刀具几何形状　geometry of the cutting tool
前角　rake angle
刀尖角　tool angle
后角　clearance angle

主偏角　side cutting edge angle
副偏角　end cutting edge angle
刃倾角　back rake angle
刀体　body
刀柄　shank
刀片　tip
可转位刀片　indexable insert tip
刀具寿命　tool life
前刀面　face

后刀面　flank
主后刀面　major flank
副后刀面　minor flank
★ 切削刃　cutting edge
主切削刃　tool major cutting edge
副切削刃　tool minor cutting edge
刀尖　corner
工件表面　workpiece surface
基面　tool reference plane
工作平面　working plane
切削平面　tool cutting edge plane
刀具总切削力　total force exerted by the tool
切削功率　cutting power
金属切除率　material removal rate
★ 切屑　chip
切屑种类　types of chips
带状切屑　continuous chips
挤裂切屑　sheared chips
崩碎切屑　discontinuous chips
剪切面　shear plane
剪切角　shear angle
已加工表面质量　quality of the machined surface
积屑瘤　built-up edge
直角切削　orthogonal cutting
斜角切削　oblique cutting
切削热　heat in cutting
切削温度　cutting temperature
后刀面磨损　flank wear
刀面磨损　face wear
月牙洼磨损　crater wear
刀具破损　tool failure
★ 切削液　cutting fluid
车刀　turning tool
可转位车刀　index turning tool
成形车刀　copying turning tool
铣刀　milling cutter
麻花钻　twist drill
中心钻　center drill
铰刀　reamer
手用铰刀　hand reamer
机用铰刀　machine reamer
镗刀　boring tool
丝锥　tap
板牙　threading die
拉刀　broach
推刀　push broach
齿轮滚刀　gear hob
蜗轮滚刀　worm gear hob
插齿刀　shaper cutter/gear shaper cutter
砂轮　abrasive wheel
齿轮刀具　gear cutting tool
高速钢　high-speed steel
硬质合金　sintered cemented carbide
碳素钢　carbon steel
合金钢　alloy steel
磨具　grinding tool
磨粒　grain
粒度　grain size
结合剂　bond
硬度　grade, hardness
组织　structure
★ 磨料　abrasive
金刚石　diamond
刚玉　alumina
黑碳化硅　black silicon carbide
绿碳化硅　green silicon carbide
立方氮化硼　cubic boron nitride
平型砂轮　straight wheel
薄片砂轮　thin grinding wheel
筒形砂轮　cylinder wheel
杯形砂轮　cup wheel
碗形砂轮　taper cup wheel
碟形砂轮　dish wheel

1 Metal cutting can be defined as a process during which the shape and dimension of a workpiece are changed by removing some of its material in the form of chips. The latter are separated from the workpiece by means of cutting tool, which must have higher hardness compared with that of the workpiece as well as certain geometric characteristics that depend upon the conditions of the cutting operation.

金属切削可以定义为通过切除金属使其成为切屑而改变工件形状和尺寸的过程。切屑通过刀具的作用从工件上被切下来，刀具相对于工件来说要有较高的硬度，并且根据加工条件要有一定的几何形状。

2 The forces acting on the tool and the workpiece are called cutting forces. In the cutting process

cutting forces directly influence the generation of cutting heat, and further influence tool wear, tool life and the quality of the machined surface.
作用在工件和刀具上的力被称为切削力。在切削过程中，切削力直接影响切削热的生成，进而影响刀具磨损、刀具寿命及已加工表面质量。

3 Cutting heat and temperature have significant influence on machining accuracy and the quality of the machined surface. Therefore, the research on cutting heat and temperature has a significant meaning.
切削热和切削温度对加工精度和已加工表面质量影响很大。因而对切削热和切削温度的研究具有重要意义。

第三节　金属切削机床及设计
Section 3　Metal Cutting Machine Tool and Design

★ 车床　　lathe
★ 铣床　　milling machine
★ 刨床　　planing machine
★ 磨床　　grinding machine
拉床　　broaching machine
齿轮加工机床　gear cutting machine
组合机床　modular machine tool
★ 加工中心　machining center
床身　　lathe bed
床头箱　headstock
尾座　　tailstock
导轨　　guideway
光杠　　feed rod
溜板箱　apron
刀架　　tool post
机械制造装备　manufacturing equipment
★ 工艺装备　tooling
机床　　machine tool
★ 通用机床　general purpose machine tool
★ 专用机床　special purpose machine tool
★ 特种加工机床　non-traditional machine tool
专门化机床　specialized machine tool
组合机床　modular machine tool
普通机床　general accuracy machine tool
精密机床　precision machine tool
高精度机床　high precision machine tool
超高精度机床　ultra precision machine tool
自动换刀装置　automatic tool changer（ATC）
自动机床　automatic machine tool
半自动机床　semi-automatic machine tool
仿形机床　copying machine tool
数控机床　numerical control machine tool/NC machine tool
主参数　main parameter
★ 运动参数　movement parameter

主轴转速　spindle speed
标准公比　standard common ratio
★ 动力参数　power parameter
★ 转速图　speed chart
有级变速　step speed changing
无级变速　stepless speed changing
滑移齿轮　slip gear
离合器　clutch
交换齿轮　changing gear
★ 三爪夹盘　three-jaw universal
★ 中心架　steady rest
★ 床身　bed
★ 底座　base
★ 导轨　slideway, guideway
★ 主轴箱　spindlehead, headstock
变速箱　gearbox
进给箱　feed box
丝杠　lead screw
★ 主轴　spindle
支承件　supporting
压板　stripe
滚动导轨　rolling guide way
立柱　column
横梁　rail/beam
刀库　tool magazine
主轴部件　spindle assembly
主轴材料　spindle material
主轴结构　spindle construction
★ 主轴　spindle
齿轮传动　gear drive
主动齿轮　driving gear
从动齿轮　driven gear
中心距　center distance
大齿轮　gearwheel
小齿轮　pinon

带传动	belt drive	润滑	lubricate
传动比	transmission ratio	润滑脂	lubricating grease
★轴承	bearing	润滑油	fluid oil
推力轴承	thrust bearing	动压轴承	hydrodynamic bearing
滑动轴承	plane bearing/sliding contact bearing	静压轴承	hydrostatic bearing
滚动轴承	rolling bearing/angular contact radial bearing	多楔滑动轴承	lobed plain bearing
向心轴承	radial bearing/single direction thrust bearing	多油楔轴承	multi-oil wedge bearing
		抗弯惯性矩	bend product of inertia
球轴承	ball bearing	抗扭惯性矩	torsional product inertia
滚子轴承	roller bearing	肋条	rib
圆锥滚子轴承	tapered roller bearing	矩形导轨	rectangle guide way
滚针轴承	needle roller bearing	三角形导轨	triangle guide way
磁力轴承	magnetic bearing	燕尾形导轨	dovetail guide way
		圆柱形导轨	column guide way

1 Most of the mechanical operations are performed on five basic machine tools: the drill press, the lathe, the shaper or planer, the milling machine, and the grinding machine.
大部分机床加工都是在五种最基本的机床上完成的:钻床、车床、刨床、铣床和磨床。

2 The oldest and most commonly used machine tool is the lathe, which removes materials by rotating the workpiece against a single point cutter.
车床是最早和最常用的机床,它通过工件与单刃刀具的相对旋转运动来切除材料。

3 Milling machines must provide a rotating spindle for the cutter and a table for fastening positioning and feeding the workpart.
铣床必须配有能使刀具旋转的主轴及用于工件定位夹紧及进给的工作台。

4 Today we are living in a society greatly affected by computer. It is still important for a student to know basic knowledge about standard machines. This knowledge will provide the necessary background for a person seeking a career in the machine tool trade.
现今我们生活的社会深受计算机的影响。但学生仍有必要学习普通机床最基本的知识。这种知识对学生将来从事机床方面的业务提供必要的背景。

5 Machining centers are multipurpose NC machines that are capable of performing a number of different machining processes at a time.
加工中心是多用途数控机床,它可以同时完成不同的加工过程。

6 Depending upon the principle of tooth-flank forming, the cutting of cylindrical gear wheels is mainly performed either by the form or generating methods.
根据齿侧成形原理,圆柱齿轮的切削主要按成形法或展成法加工。

7 After completing this section, students will be able to:
(1) list the operations which can be performed on a lathe.
(2) know how the size of a lathe is designated.
(3) name the four main units of a lathe.
(4) state the purposes of the following:
spindle
lead screw and feed rod
quick-change gearbox
split-nut
完成本节学习之后,学生应能:
(1) 列举出车床上可以进行的操作。
(2) 知道车床的型号是由哪个参数决定的。

第六章 机械制造技术基础及装备设计

(3) 说出车床的四个主要组成部分。
(4) 指出下列元件的作用：
主轴
丝杠和光杠
进给箱
开合螺母

8 The lathe bed is the main frame, involving a horizontal beam on two vertical supports.
床身是主要框架，包括两根立柱上的横梁。

9 The headstock is fixed at the left hand side of the lathe bed and includes the spindle whose axis is parallel to the guideways (the slide surface of the bed).
床头箱固定在车床的左手边，其中包含与导轨(床身滑动面)相平行的主轴。

第四节 机床夹具原理与设计

Section 4 Principle and Design of Jigs and Fixtures

★ 自由度　freedom	刨床夹具　fixture for planing machine
找正　mark out	随行夹具　workholding pallet
★ 夹紧　clamp	定位件　locating piece/locating element
对刀块　setting block	夹紧件　clamping element
定向键　tenon	导向件　guiding element
圆柱销　full diameter location pin	对刀件　element for aligning tool
削边销　flatted location pin	插床夹具　fixture for slotting machine
夹盘　chuck	磨床夹具　fixture for grinding machine
夹紧力　clamping force	★ 夹紧件　clamping element
★ 夹具　fixture, jig	导向件　guiding element
专用夹具　special fixture	★ 对刀件　element for aligning tool
通用夹具　universal fixture	支承板　strip
组合夹具　modular jig and fixture	气动夹盘　pneumatic chuck
可调夹具　adjustable fixture	液压夹盘　hydraulic chuck
成组夹具　modular fixture	夹头　collet chuck
标准夹具　standard fixture	电磁吸盘　electromagnetic chuck
手动夹具　manual fixture	平口虎钳　plane-jaw vice
气压夹具　pneumatic fixture	夹紧力　clamping force
液压夹具　hydraulic fixture	芯轴　arbor
电动夹具　electric fixture	弹性夹头　collet
机床夹具　machine tool fixture	虎钳　vise
车床夹具　lathe fixture	分度头　indexing head/dividing head
铣床夹具　fixture for milling machine	钻套　drill bush
镗床夹具　fixture for boring machine	工作台　worktable
钻床夹具　fixture for drilling machine	

1 A drill jig is a device to ensure that a hole could be drilled, tapped, or reamed in a workpiece or will be machined in the proper place. If the operations includes machining operations like milling, planning, shaping, turning, etc., the term fixture should be used.
钻模是用于保证在工件的正确位置上钻孔、攻丝、铰孔的装置。如果加工是针对铣削、刨削及车削等，就要用到夹具。

2 The location element contacts with the datum surface of the workpiece, and is used to determine the correct position of the workpiece in the jig or fixture.

定位元件与工件的定位基面接触,用于确定工件在钻模和夹具上的正确位置。

3 Clamping device is used to clamp the working. In other words, it secures the workpiece in its original correct position during machining.

夹紧装置用于夹紧工件。换句话说,它用于保证工件在加工过程中仍保持原来的正确位置。

4 A fixture is a work-holding device fastened to the table of a machine or to a machine accessory, such as a rotary table. It is designed to hold workpieces that cannot be readily held in a vise or in production work when large quantities are to be machined. The fixture must be designed so that the identical parts, when held in the fixture, will be positioned exactly and held securely.

夹具是工件夹持装置,固定在机床工作台或附件上,如旋转台。当大量生产虎钳无法夹持住工件时,就要设计夹具。设计的夹具加工相同的零件时要定位准确,夹紧可靠。

5 A fixture holds the work during machining operations but does not contain special arrangements for guiding the cutting tool, as drill jigs do.

夹具在机器加工时夹持工件,但与钻模不同,没有专门引导刀具的装置。

第五节　加工质量分析与控制
Section 5　Analysis and Control for Machining Quality

已加工表面　machined surface	珩磨　honing
表面质量　surface quality	抛光　polishing
表面粗糙度　surface roughness	内应力　internal stress
★加工精度　machining accuracy	系统性误差　systematic error
★加工误差　machining error	随机误差　random error
工艺　technology	正态分布　normal school
★机械制造工艺　manufacturing technology	概率　probability
表面冷作硬化　cold hardening of the surface layer	冷作硬化　cold deformation strengthening
原材料　raw material	疲劳强度　fatigue strength
毛坯　blank	残余应力　residual stress
几何误差　geometrical error	金相组织　metallographic structure
导轨　guide way	振动　vibration
定位误差　positioning error	强迫振动　forced vibration
刚度　rigidity	自激振动　self-excitation vibration
热源　heat source	自由振动　free vibration

1 Not only the magnitude of surface roughness but also the profile shape and the texture direction of surface have an influence on wear characteristics.

不仅表面粗糙度的大小对磨损特性有影响,表面轮廓形状及纹理方向对磨损特性也有影响。

2 The manufacture of dimensionally accurate, closely fitting parts is essential to interchangeable manufacture. The accuracy and wearability of mating surfaces are directly proportional to the surface finish produced on the part.

对于制造的互换性来说,至关重要的是尺寸精确,拟合密切。配合表面的精度和耐磨性与零件加工表面的质量成正比。

3 Cold harding of the surface layer can decrease the parts wear remarkably.

表面层的冷作硬化可以大大减少零件的磨损。

4 The forming of machined surface roughness results from three aspects: a geometric factor, a physical factor and vibration of the technological system.

表面粗糙度由三方面的原因形成:几何因素、物理因素及系统的振动。

第六章 机械制造技术基础及装备设计

第六节 工艺规程设计
Section 6　Process Planing

工艺规程	process rule	工艺凸台	false boss
工艺卡片	process card	加工余量	machining allowance
★粗基准	rough location feature	工序余量	operation allowance
★精基准	finished location feature	切入量	approach
尺寸链	dimensional chain	切除量	over travel overrun
★工序	operation	光整加工	finishing cut
工步	step	超精加工	superfinishing
工位	position	抛光	polishing
★安装	setup	成组技术	group technology (GT)
工作行程	working stroke	装配精度	assembly precision
空行程	idle stroke	装配方法	assembly method
工艺孔	auxiliary hole	装配误差	assembly error
中心孔	center hole		

1 It is important that the sequence of operations should be carefully planned in order to produce a part quickly and accurately. Improper planning or following a wrong sequence of operations often results in spoiled work.
为了快速而精确地加工零件必须精心安排加工工艺顺序(流程),否则将损伤零件加工表面。

2 The job performed in one position is called one operation.
工件在所占据的每一个位置上完成的加工叫做一个工位。

3 Operations with multi-positions can enhance the production efficiency and the relative position precision among the manufacturing surfaces.
多工位加工可以提高生产率及加工表面间的相对位置精度。

第七节 物料输送系统及仓储装置设计
Section 7　Material Handling System and Warehouse Design

★仓储装置	warehouse	机器人	robot
★上下料装置	loader and unloader	运输小车	vehicle
★物料输送系统	material handling system	交换工作台	pallet changer
料斗	hopper	高层货架	high-rise rack
机械手	manipulator	堆垛机	stacker crane

1 The goal of an automated storage and retrieval system is to deliver the right material to the right place at the right time.
自动存取系统的作用就是要在恰当的时间将恰当的物品送到恰当的地点。

2 Material is held in storage and then issued to the point of use as close to the time of use as possible.
物料先存在仓库中然后在尽可能接近使用的时间被送到使用地点。

3 Some FMSs are configured with automated guided vehicles (AGVs) as a principal means of materials handling.
一些柔性制造系统配备有自动输送小车作为主要的物料输送工具。

第八节 机械加工生产线
Section 8　Machining Production Line Design

★自动生产线	automatic production line		machine
组合机床自动生产线	transfer line of modular machine/automatic production line of modular	★机动时间	mechanic operating time
		★辅助时间	auxiliary motion time

加工示意图　process sketch chart
工作循环　work cycle
工作行程　work travel
布局　layout
柔性　flexibility

柔性制造系统　flexible manufacturing system（FMS）
柔性加工单元　flexible manufacturing cell（FMC）

1 The touch trigger probe provides a signal to allow the "latching" of the CMM coordinates at the moment of probe triggering.
接触触发式探头在触头触发的瞬时提供一个信号,使坐标测量机锁存下此瞬间的坐标值。

2 If the workpiece (or several different workpieces on a pallet) passes inspection, the pallet is returned to the CMM's shuttle position for AGV transporting to the queuing carrousel to await part removal or refixture for a subsequent machining operation.
如果工件(或一个托盘上若干个不同的工件)检查合格,则该托盘便回到坐标测量机的穿梭交换位置,由 AGV 将其运送至排队转台,等待卸下工件或进行重新装夹,以进入随后的机械加工。

第九节　先进制造技术
Section 9　Advanced Manufacturing Technology

★超精密加工　superfinishing
超高速切削　superhigh speed cutting
制造哲理　manufacturing philosophy
生产模式　production mode
计算机集成制造系统　computer integrated manufacturing system（CIMS）
管理信息系统　management information system（MIS）
技术信息系统　technological information system（TIS）
自动化制造系统　manufacturing automation system（MAS）
质量保证系统　computer aided quality assurance（CAQ）
计算机辅助工艺规程设计　computer aided process planning（CAPP）
物料清单　bills of material（BOM）
工艺路线　process routing
工作中心　working center
主生产计划系统　master production scheduling system（MPSS）
★并行工程　concurrent engineering（CE）
★虚拟制造　virtual manufacturing（VM）
快速原型制造　manufacturing rapid prototype（MRP）
快速模具制造　manufacturing rapid tooling（MRT）
快速精铸　quick casting
★快速求反工程　quick reverse engineering

1 The introduction of advanced manufacturing technology has made it possible to machine new space-age materials and to produce shapes which were difficult or often impossible to produce by other methods.
有了先进的制造技术,加工新型航空材料和加工其他方法无法加工或很难加工的形状成为可能。

2 Manufacturing automation protocol (MAP) is a communication standard developed to promote compatibility among automated manufacturing systems produced by different vendors. It allows different machines to talk to each other.
制造自动化协议(MAP)是为了实现由不同商家提供的自动化制造系统间的兼容而开发出来的通信协议,可完成不同机器之间的对话。

3 The scheduler function involves planning how to produce the current volume of orders in the FMS, considering the current status of machine tools, work-in-process, tooling, fixtures and so on.
生产调度程序的功能涉及如何用 FMS 生产方式按当前定货量来制造产品,在安排计划时要同时考虑到机床的现有状态、加工过程、工模具和夹具等。

第七章　微机原理与应用
Chapter 7　Microcomputer Principle and Application

第一节　计算机基础知识
Section 1　Basic Knowledge of Computer

小型计算机	minicomputer	十进制	decimal system
微型计算机	microcomputer	十六进制	hex/hexadecimal system
计算机	computer	原码	original code
数	number	反码	base minus ones complement
编码系统	coding system	补码	base complement
集成电路	integrate circuit	硬件	hardware
二进制	binary system	软件	software
八进制	octal system		

1 Computers are divided into analogue computer and digital computer.
计算机被分为模拟计算机和数字计算机。

2 One of the basic jobs of computer is the processing of information.
处理信息是计算机的基本工作之一。

3 For this reason, computers can be defined as very-speed-electronic devices, which accept information in the form of instructs called a program and characters called data, perform mathematical and/or logical operations on the information, and then supply results of these operations.
由于这个原因,计算机被称作是一种能够接受命令和数据,并能对这些信息进行数字或逻辑运算,最终输出运算结果的高速电子装置。

第二节　MCS-51 单片机概述
Section 2　Microcomputer Overview

中央处理器	central processing unit	字节	byte
单片机	single chip processor	字	word
4 位机	4-bit computer	双字	double word
8 位机	8-bit computer	主频	primary frequency
16 位机	16-bit computer	时钟	clock
位	bit	指令周期	instruction cycle

1 The 8051 series of micro controllers is 8-bit computer.
8051 系列微处理器是八位的计算机。

2 Bit is the least unit to store for 8051 micro controller.
位是 8051 单片机进行存储的最小单元。

3 The 8051 series of micro controllers are highly integrated single chip microcomputers with an 8-bit CPU, memory, interrupt controller, timers, serial I/O and digital I/O on a single piece of silicon.
8051 系列单片机是一个高集成度的微处理器,在其硅片上有八位 CPU、存储器、中断控制器、定时器及串行和并行接口。

第三节　单片机的结构原理
Section 3　MCU Structure Principle

集成芯片　integrated chip
单片机　single-chip microcomputer
引脚　pin
振荡器　oscillator
程序存储器　program memory
数据存储器　data memory
控制电路　control circuit
I/O 端口　input and output port
定时器　timer
计数器　counter
中断源　interruption source
串行口　serial port
并行口　parallel port
数据总线　data bus
地址总线　address bus
控制总线　control bus
运算器　arithmetic organ
累加器　accumulator
寄存器　register
程序状态字　program status word
进位标志　carry flag
辅助进位标志　auxiliary carry flag
溢出标志　overflow flag
零标志　zero flag

符号标志　sign flag
控制器　controller
译码器　encoder
程序计数器　program counter
位寄存器　bit register
字节寄存器　byte register
工作寄存器　working register
缓冲器　buffer
数据缓冲器　data buffer
负载能力　load capacity
读　read
写　write
时序　sequence
脉冲　pulse
复位　reset
置位　set
掉电　power down
堆栈　stack
数据区　data area
程序区　program area
特殊功能寄存器　special function register
位地址　bit address
字节地址　byte address
访问　access

1 The 8051 has a fairly complete set of arithmetic and logical instructions.
8051 有一套非常完善的算术、逻辑运算指令。

2 Arithmetic and logical device has an 8×8 multiplication and an 8/8 division.
算术、逻辑运算器包括 8×8 乘法和一个 8/8 除法。

3 The 8051 is particularly good at processing bits (sometimes called Boolean Processing). Using the carry flag in the PSW as a single bit accumulator, the 8051 can move and do logical operations between the Bit Memory space and the carry flag.
8051 在位处理方面非常出色。标志寄存器中的进位标志寄存器作为位累加器，8051 可以在进位标志和位存储器中进行传送和逻辑操作。

第四节　MCS-51 指令系统
Section 4　MCS-51 Instruction System

绝对调用　absolute call
加法指令　add
带进位加法指令　add with carry
绝对跳转　absolute jump
逻辑与　logical and

比较转移　compare jump
清零　clear
取补　complement
十进制调整　decimal adjust
减指令　decrement

第七章 微机原理与应用

除指令	divide	入栈	push stack
循环控制指令	loop control instruction	返回	return
加指令	increment	中断返回	return from interrupt
位置位跳转	jump if bit set	循环左移	rotate left
位置位跳转并清零	jump & clear bit if bit set	带进位位循环左移	rotate left thru carry
进位位置位跳转	jump if carry set	循环右移	rotate right
跳转	jump	带进位位循环右移	rotate right thru carry
位复位跳转	jump if bit not set	置位	set bit
进位位复位跳转	jump if carry not set	短跳转	short jump
累加器为零跳转	jump if accumulator zero	带借位位减法	subtract with borrow
长调用	long call	累加器高低位交换	swap nibbles
长跳转	long jump	字节交换	exchange bytes
传送	move	数据交换	exchange digits
机器码传送	move code	异或	exclusive or
外部数据传送	move external	调用	generic call
乘法指令	multiply instruction	右移	shift right
空操作	no operation	左移	shift left
或指令	inclusive OR	逻辑非	logical negation
出栈	pop stack		

1. Instruction is the basic unit of computer softwares.
   指令是计算机软件的基本单元。
2. Program control instructions include the usual unconditional calls and jumps.
   程序控制指令包括通常的无条件调用和转移指令。
3. Program control instructions also include conditional relative jumps based on the carry flag, the accumulator's zero state, and the state of any bit in the Bit Memory space.
   程序控制指令也包括基于进位标志、零标志和位存储区中任何位的条件相对转移。

第五节 程序设计
Section 5 Program Design

指令	instruction	目标文件列表	object file listing
符号	symbol	汇编程序文件	assembler file
标号	label	最小系统	minimum system
汇编程序	assembler	最大系统	maximum system
汇编控制	assembler control	执行汇编程序	running the assembler
汇编命令	assembler directive	宏定义	macro definition
位寻址	bit addressing	宏操作	macro operators
字节寻址	byte addressing	汇编错误码	assembler error code
注释	comment	错误信息	error message
计数器	counter	保留符号	reserved symbol
句法	syntax	汇编符号集	assembler character set
操作符	operator	八位机	8-bit machine
源文件	source file	常数	constant
源文件列表	source file listing	变量	variable
目标文件	object file	缓冲器	buffer

起始地址 beginning-address	定义空间伪指令 reserve bytes of data memory
常数表 constant-table	定义字节伪指令 store byte values in program memory
变量表 variable table	
赋值伪指令 define symbol	定义字伪指令 store word values in program memory
数据地址赋值伪指令 define internal memory symbol	
	起始地址伪指令 set segment location counter
位地址赋值伪指令 define internal bit memory symbol	汇编结束伪指令 end of assembly language source file

1 Program designing is one of the important contents for computer applications.
程序设计是计算机应用的重要内容之一。

2 The 8051 assembler can create a machine language object file.
8051 汇编程序可以产生一个机器语言目标文件。

3 Object file can be used to program EPROM, or used to produce mask ROM.
目标文件可以用来对 EPROM 进行编程,或用来生产掩膜 ROM。

第六节　硬件基础
Section 6　Hardware Basics

接口 interface	边沿触发 edge trigging
接口电路 interface circuit	入口地址 entry address
可编程接口 programmable interface	复位状态 reset state
优先级 priority level	存储器 memory
中断 interrupt	存储器扩展 memory expanding
中断源 interrupt source	键盘 keyboard
中断请求 interrupt request	显示器 display unit
响应 response	数模转换 digital to analog conversion
触发 trigging	模数转换 analog to digital conversion
电平触发 level trigging	触发器 trigger

1 CPU is the abbreviation of central processing unit.
CPU 是中央处理单元的缩写。

2 The CPU is the most important component of a computer, and is called computer's brain.
CPU 是硬件系统最重要的部分,被称为计算机的大脑。

3 The function of CPU is to extract instruction stored in the main memory by fetching and examining them, then executing them one after another.
CPU 的功能是取出和检查存储在主存储器中的指令并一条一条地执行指令。

第八章 冷冲压模具设计
Chapter 8　Cold Stamping Die Design

第一节　冲裁模具设计基础
Section 1　Punching Die Design Basic

模具　die/mould
冲压　punch
压力机/冲床　punch press
冲压过程　punch process
模具设计　mold design
冷冲压模具　cold stamping die

冲压工艺　stamping process
工序划分　processes defined
分离工序　separation process
成型工序　forming process
冲压工艺参数　stamping process parameters

1. plasticity in the suppression techniques, such as stamping, bending, flanging, tensile plate
   适用于可塑性的压制工艺，如冲压、弯曲、翻边、薄板拉伸等
2. integration & application of multi-objective optimization for stamping process of auto-body panels
   车身覆盖件冲压工艺多目标优化系统的集成与应用
3. research for the automatic design system of punching dies
   冲压模具自动设计系统的研究

第二节　冲裁工艺与冲裁模设计
Section 2　Blanking Process and Blanking Die Design

落料　blanking
外形落料　appearance blanking
切断　cut off
切舌定距　stamping and bending fixed distance
排样设计　layout design
切边　trimming
模具的工作过程　die working process

导柱　guide post
导向　guiding
导向板　guide plate
导套　guide bush
冲孔　hole punching
冲裁模　blanking die

1. research and implementation of the stamping force calculation in the blanking die CAD system
   冲裁模CAD中冲压力计算的研究与实现
2. calculation formulas of edge size in a blanking die.
   冲裁模刃口尺寸的计算公式
3. failure analysis of die and countermeasures
   冲裁模的失效分析与对策

第三节　弯曲工艺与弯曲模具设计
Section 3　Bending Process and Bending Die Design

弯曲模　bending die
弯曲工艺　bending process
相对弯曲半径　relative bending radius
回弹现象　rebound phenomenon
中心线弯曲角　camber angle
回弹值　rebound value
浮动式凹模　floating cavity

曲面凹模　curved die
弯曲测试　bending test
弯曲力　bending force
弯曲半径　bending radius
弯曲模结构　bending die structure
多次弯曲　multi-step bending

[1] bending process and die design of thick wall support
厚壁支架的弯曲工艺及模具设计
[2] process parameter for the tension leveling of metal strip with the finite element method
基于有限元方法的金属板材连续拉伸矫直工艺参数
[3] theory and experiment on elastic-plastic bending of wide plates
宽板弹塑性弯曲工艺的理论与实验

第四节 拉深工艺与拉深模具设计
Section 4 Stretching Process and Stretching Die Design

拉深 drawing	拉应力区 tension stress zone
拉深模 drawing die	断裂拉伸应力 tensile stress at break
深度拉伸塑性变形 deep drawing plastic deformation	抗拉强度 tensile strength
	体积不变原则 the principle of constant volume
拉深成形 drawing forming	相似原则 principle of similarity
变形现象 metamorphism	多次拉深 multi-stage deep drawing
凸缘变形区 flange deformation region	

[1] the load curve of critical wrinkles variable blank holder force in the process of deep drawing cylinder parts
圆筒件拉深成形临界防皱变压边力加载曲线
[2] analysis of nonaxisymmetric phenomenon of titanium alloy cupulate specimen in drawing process
钛合金杯形件拉深成形非轴对称现象的分析
[3] in multi-stage drawing process, the total amount of deformation is divided into several parts, and formed sequentially in several drawing steps.
在多阶段拉深过程中，总变形量被分成几部分，从而形成了几个拉深的步骤。

第五节 模具加工方法与刀具
Section 5 Mould Process Method and Tool

模具加工方法与刀具 mould process method and tool	车刀 turning tool
钻削 drilling	可转位车刀 index turning tool
车削 turning	成形车刀 copying turning tool
铣削 milling	铣刀 milling cutter
磨削 grinding	麻花钻 twist drill
铰削 reaming	中心钻 center drill
镗削 boring	铰刀 reamer
电火花加工 electro-discharge machining (EDM)	镗刀 boring tool
电解加工 electrochemical machining	丝锥 tap
激光加工 laser beam machining	板牙 threading die
超声波加工 ultrasonic machining	拉刀 broach
	推刀 push broach
	砂轮 abrasive wheel

[1] NC machining process analysis of typical mould molding parts
典型模具成型零件数控加工工艺分析
[2] system for tool path generation of 2D contour milling of die
模具二维轮廓铣削加工刀具轨迹生成系统
[3] mirror face processing method of precise & big area automobile mould
大面积精密汽车模具镜面加工方法的研究

第八章 冷冲压模具设计

第六节 模具加工机床与夹具
Section 6　Mould Machine Tools and Fixtures

机床	machine tool	分度头	indexing head/dividing head
车床	lathe	钻套	drill bush
铣床	milling machine	工作台	worktable
刨床	planing machine	自由度	freedom
磨床	grinding machine	找正	mark out
拉床	broaching machine	夹紧	clamp
齿轮加工机床	gear cutting machine	对刀块	setting block
组合机床	modular machine tool	定向键	tenon
数控机床	numerical control machine	圆柱销	full diameter location pin
加工中心	machining center	削边销	flatted location pin
通用机床	general purpose machine tool	夹盘	chuck
专用机床	special purpose machine tool	夹紧力	clamping force
特种加工机床	special machine	夹具	fixture/jig
虎钳	vise		

1 Punch press: Machine tool that changes the size or shape of a piece of material, usually sheet metal, by applying pressure to a die in which the workpiece is held.
  冲床：通过将压力加到装有工件的模具上，以改变一片材料，通常是金属片大小和形状的机床。

2 We use reasonable machining process to greatly improve the machining precision of the rolling dies, and reduce the times of trying dies.
  采用合理的机加工工艺，可大大提高模具加工精度，减少试模次数。

3 A fixture holds the work during maching aperations but does not contain special arrangement for guiding the cutting tool, as drill jigs do.
  夹具在机加工时夹持工件，但与钻模不同的是，它没有专门引导刀具的装置。

4 The affected cause of machining accuracy and surface roughness is diverse part in WEDM for die & mould machining
  在电火花线切割加工中，影响模具加工精度与表面粗糙度的原因是多方面的。

第九章 计算机辅助绘图
Chapter 9　AutoCAD

第一节　AutoCAD 概述
Section 1　Introduction of AutoCAD

计算机辅助设计　computer-aided design
通用辅助设计系统　common use aided design system
二维图形　two-dimension patterning
三维实体　three-dimensional substance
图形编辑　figure edit
智能化　intelligentize
标题栏　headline
菜单栏　menu
工具栏　tool menu
命令行　order line
状态栏　status
快捷菜单　shortcut menu
图层　layer
样板　sample plate
对象选择　object selection
二次开发　secondary development
基本功能　basic functions
选项卡　option
文件加密　file encryption
设置参数　set the parameter
绘图界限　drawing boundaries
公制　metric system
英制　imperial system
打印预览　print preview

1 AutoCAD is a design software developed specifically for computer graphics.
AutoCAD 是专门为计算机绘图开发的设计软件。

2 As the engineering technology, computer-aided design has shown a strong technical strength in designing, drawing and mutual collaboration.
计算机辅助设计作为工程领域的主要技术,在设计、绘图、相互协作方面显示出了强大的技术实力。

第二节　绘图命令栏
Section 2　Drawing Command Bar

点　point
直线　line
多段线　polyline
多线　multi-line
射线　ray
圆　circle
圆弧　arc
圆环　ring
椭圆　ellipse
面域　surface domain
多行文本　multiple lines of text
三角形　triangle
矩形　rectangle
正多边形　regular polygon
样条曲线　spline curve
块　block
块定义　block definition
插入块　block is inserted
填充　filling

1 In order to facilitate the work of drawing, CAD provides some control graphic display command; under normal circumstances, these commands can change the graphics on the screen display, and graphics operator is according to the desired position, and the proportion of display range.
为了便于绘图工作,CAD 提供了一些控制图形显示的命令。一般情况下,这些命令能改变图形在屏幕上的显示方式,使图形按操作者期望的位置、比例和显示范围显示。

2 The utility of adjuvant tool command does not create or edit entity, but the utility of these commands can build a better drawing environment, so that users can use less time to draw the high-precision graphics.

实用的辅助工具命令并不能生成或编辑实体，但使用这些命令可以建立一个更好的绘图环境，使用户能够用较少的时间绘制出高精度的图形。

第三节　编辑命令栏
Section 3　Edit Command Bar

复制	copy	拉伸	tensile
镜像	mirror	直线拉长	straight-line stretched
阵列	array	比例缩放	scaling
偏移	offset	打断	break
旋转	rotation	倒角	chamfer
移动	move	倒圆角	rounding
删除	delete	多段线编辑	multi-line editing
修剪	trim	修改文本	modify the text
延伸	extend		

1 The user can conveniently draw horizontal, vertical line with orthogonal function; object capture can help user pick up special points on geometry, and tracking function to draw an oblique line and along different direction positioning point becomes easier.

正交功能使用户可以很方便地绘制水平、竖直直线，对象捕捉可帮助用户拾取几何对象上的特殊点，而追踪功能使画斜线及沿不同方向定位点变得更加容易。

2 CAD has a powerful editing functions; objects can be moved, copied, rotation, stretching, extend, array, pruning, zooming.

CAD 具有强大的编辑功能，可以移动、复制、旋转、阵列、拉伸、延长、修剪、缩放对象。

第四节　辅助绘图命令
Section 4　Aided Drawing Commands

属性匹配	attributes match	输入文件	input file
设置颜色	set the color	自定义设置	custom settings
图层操作	layer operation	重命名	rename
线形	linear	捕捉栅格	capture grid
线形比例	linear proportion	正交功能	orthogonal function
线宽	linewidth	对象捕捉	object capture
图形单位	graphics unit	极轴追踪	axis tracking
属性定义	attribute definition	命名视图	named view
编辑属性	edit properties	文字样式	text style
对齐	alignment	修改特性	modify the characteristic
退出	quit		

1 Master "Scaling" commands and object selection methods and techniques are one of the most important basics in AutoCAD drawing.

熟练掌握"缩放"命令和选择对象的方法和技巧是 AutoCAD 绘图的重要基础之一。

2 While editing, you need to choose the edited object.

在编辑的时候需要选择被编辑的对象。

3 The graphics except freeze, locked layer on top of all the objects are selected using all the way choice.

将图形中除冻结、锁定层上的所有对象选中可以使用全部方式选择对象。

第五节 标注命令
Section 5　Marking Commands

直线标注	a straight line marked	连续标注	continuous marking
对齐标注	alignment marked	标注样式	marking style
半径标注	radius marked	编辑尺寸	edit dimension
直径标注	diameter marked	标注样式管理器	dimension style manager
角度标注	angular marked	标注特征比例	marked characteristic ratio
中心标注	center marked	舍入精度	rounding precision
坐标标注	coordinate marked	起点偏移量	starting point offset volume
折弯标注	bending marked	极限偏差	limit deviations
调整间距	adjust the spacing	极限尺寸	limit size
标注形位公差	marking geometric tolerance	基本尺寸	basic dimensions
快速引出标注	quickly leads to marked	替换系统标注变量	replace the system marked variable
基线标注	baseline marked		

1 Click the select dimension style manager on the left; select the style you want to modify, and then click on the right side of the edit button can modify based on the demand.
点击左面选择标注样式管理器,选择你想修改的样式,再点击右面的修改按钮就可以根据需求进行修改。

2 Dimension style setting depends entirely on the industry standard and drawing standard.
标注样式的设置完全取决于行业标准和绘图标准。

3 General digital should note in the dimension line above, may also note in the dimension line interruption.
一般数字应注在尺寸线的上方,也可注在尺寸线的中断处。

第六节 综合运用各种绘图方法绘制工程图
Section 6　Comprehensive Use of Various Drawing of Engineering Drawing

零件	component	面部曲线	face curve
建造模型	building model	吸收	absorbed
集体特征	the collective identity	对齐	align
凸凹特征	convex and concave characteristics	定位点	anchor point
切除特征	removal characteristics	注解	annotation
抽壳特征	shell features	外观标注	appearance callout
阵列特征	pattern features	区域剖面线	area hatch
装配体	assembly	附加点	attachment point
智能尺寸	smart dimension	基准轴	axis
工程图	drawing	基准尺寸	baseline dimension
装配体配合	assembly mate	零件序号	balloon
筋	rib	基体	base
拔模	draft	边界框	bounding box
子装配体	sub-assembly	中心线	centerline
自动过渡	automatic transition	解除爆炸	collapse
草图捕捉	sketches capture	分离工程图	detached drawing
快速捕捉	quickly capture	局部视图	detail view
交叉曲线	cross curve	爆炸视图	exploded view

第九章　计算机辅助绘图

1 In the new project plan, through the model set up and use can improve mapping efficiency.
在新建工程图时,通过样板的设置和调用能很好地提高作图效率。

2 In the three-dimensional map conversion into two-dimensional engineering drawings, CAD and Pro/E and SolidWorks are specific interface conversion.
在由三维图转换为二维工程图时,CAD 和 Pro/E 及 SolidWorks 之间都有特定的接口实现相互转换。

第七节　Pro/E
Section 7　Pro/E

模型树	pattern tree	着色	shading
缺省模型	default model	使用父视图造型	use parent view style
指定模板	specify template	显示剖面线	show x-hatching
空格式	empty with format	从动环境	follow environment
定向	orientation	面组隐藏线移除	hidden line removal for quilts
纵向	portrait	焊件剖面显示	weldment x-section display
横向	landscape	模型栅格	model grids
选取实例	select instance	相对坐标	relative cords
表	table	绝对坐标	absolute cords
布局	layout	注释绘图	annotation draw
设置模型	set model	审阅绘图	inspect draw
模型显示	model display	发布绘图	release draw
绘图树	sketch tree	渲染	apply colours to drawing
页面	sheet		

1 In the absence of user confirmation, the case "sketch" can remove the size or constraint called "weak size" or "weak constraint".
在没有用户确认的情况下,"草绘器"可以移除的尺寸或约束称为"弱尺寸"或"弱约束"。

2 All draft primitives can be found under the sketch menu command or the on right side of the screen tool to find the corresponding shortcut icon.
所有草绘图元都可以在 sketch 菜单下找到命令或在屏幕右边的工具条中找到对应的快捷图标。

3 In many cases, identifying the primitive shapes and sizes can use the constraint methods and can also use tagging method, and this time it will be based on the designer's intention to determine.
很多情况下,确定图元的形状和尺寸都可以用约束的方法,也可以用标注的方法,这个时候就要根据设计者的意图来确定了。

第八节　SolidWorks 的基本功能
Section 8　The Basic Function of the SolidWorks

型腔	cavity	系列零件设计表	design table
中心符号线	center mark	拔模	draft
闭环轮廓	closed profile	从动尺寸	driven dimension
碰撞检查	impact checking	驱动尺寸	driving dimension
约束	constraint	动态间隙	dynamic clearance
构造几何体	construction geometry	拉伸	extrude
坐标系	coordinate system	掏空	hollow
装饰螺纹线	cosmetic thread	干涉检查	interference detection
曲率	curvature	尺寸链	ordinate dimensions
自由度	degrees of freedom	路径	path

重建模型	rebuild	临时轴	temporary axis
相对视图	relative view	薄壁特征	thin feature
实体扫描	solid sweep	自上而下的设计	top-down design
切线弧	tangent arc	焊件	weldment
分割线	split line	斑马条纹	zebra stripe
子装配体	sub-assembly		

1 SolidWorks can provide different designs, reduce the errors in the design process and improve the quality of products.

SolidWorks 能够提供不同的设计方案,减少设计过程中的错误以及提高产品质量。

2 SolidWorks provides a completed detailed engineering drawings and recognition workshop. Engineering drawings are all related, and when you modify the drawings, 3D model, each view, assembly will updated automatically.

SolidWorks 提供了生成完整的、车间认可的详细工程图的工具。工程图所有操作都是相关的,当你修改图纸时,三维模型、各个视图、装配体都会自动更新。

第十章 互换性与技术测量
Chapter 10 Interchangeability and Techno-meterage

第一节 圆柱公差与配合
Section 1 Tolerance and Fit of Column

★ 互换性　interchangeability
★ 标准化　standardization
优先数系　series of preferred numbers
孔　hole
轴　shaft
尺寸公差　size tolerance
尺寸　size/dimension
★ 基本尺寸　basic size
★ 实际尺寸　actual size
局部实际尺寸　actual local size
极限尺寸　limits of size
★ 最大极限尺寸　maximum limit of size
★ 最小极限尺寸　minimum limit of size
最小实体极限　least material limit (LML)
最大实体极限　maximum material limit (MML)
偏差　deviation
极限偏差　limit of deviation
★ 上偏差　upper deviation
★ 下偏差　lower deviation
基本偏差　fundamental deviation
实际偏差　actual deviation
★ 零线　zero line
★ 公差带　tolerance zone
★ 公差带代号　tolerance zone symbol

★ 公差等级　tolerance grade
标准公差　standard tolerance
标准公差等级　standard tolerance grade
★ 配合　fit
★ 配合表面　mating surface
配合公差带　fit tolerance zone
★ 间隙　clearance
★ 过盈　interference
★ 间隙配合　clearance fit
★ 过盈配合　interference fit
★ 过渡配合　transition fit
最大间隙　maximum clearance
最大过盈　maximum interference
最小间隙　minimum clearance
最小过盈　minimum interference
基准孔　basic hole
基准轴　basic shaft
基孔制配合　(ISO) hole-basis system of fits
基轴制配合　(ISO) shaft-basis system of fits
基准制　benchmark system
配合公差　variation of fit; fit tolerance
优先配合　preferred fit
常用配合　commonly used fit

Tolerance of size and fit are main content of precision design.
尺寸公差与配合是精密设计的主要内容。

第二节 长度测量基础
Section 2 Basic on Length Measure

★ 测量　measure
测量对象　measure object
测量单位　measure unit
测量方法　measure method
测量精确度　measure precision
尺寸传递　dimension transfer
测量器具　measure implement
标准量具　standard measure

极限量具　limit measure
检验夹具　check up clamp
测量仪器　measuring instrument
游标式量仪　cursor measuring instrument
机械式量仪　mechanical measuring instrument
光学机械式量仪　photics mechanism measuring instrument
电动式量仪　electric measuring instrument

光电式量仪　photoelectricity measuring instrument
专用量仪　special purpose measuring instrument
通用量仪　universal measuring instrument
测量装置　measuring apparatus
三维测头　three dimensional probe/3D probe
立式测长仪　vertical comparator
万能测长仪　universal comparator
激光测长仪　laser length measuring machine
万能测齿仪　universal gear tester
渐开线检查仪　involute tester
电触式比较仪　electric-contact comparator
立式光学比较仪　vertical optical comparator
卧式光学比较仪　horizontal optical comparator
卧式阿贝尔比较仪　horizontal Abbe comparator
光电测扭仪　photoelectric torquemeter
光学测距仪　optical range finder
光学扭簧测微仪　spring-optical measuring head optical micrometer
杠杆式测微仪　lever-type micrometer
电感式测微仪　inductive gage
电容式测微仪　capacitance gage
杠杆齿轮测微仪　lever and gear type micrometer
比较仪　comparator
齿轮双面啮合检查仪　double flank gear rolling tester
齿轮径向跳动检查仪　gear radial runout tester
平直度测量仪　flatness measuring instrument
圆度仪　roundness measuring instrument
水平仪　leveling instrument/level
轮廓仪　profilometer
磁力测厚仪　magnetic thickness tester
多功能仪器　multiple function apparatus
检测装置　detection device
光电式检测装置　photoelectric detection device
自动检测装置　automatic measuring device
自动检测系统　automatic test system
绝对测量　absolute measurement
相对测量　relative measurement
接触测量　contact measurement
非接触测量　non-contact measurement
综合测量　composite measurement

单项测量　analytical measurement; single element measurement
静态测量　static state measurement
主动测量　active measurement/measurement for active control
被动测量　passive measurement
测量力　measuring force
可靠性　dependability
量具　measuring tool
单值量具　single-value measuring tool
多值量具　multi-value measuring tool
独立量具　independent measuring tool
成套量具　complete set of measuring tool
游标量具　vernier measuring tool
钢尺　steel rule
钢卷尺　steel tape
标尺　scale
游标卡尺　vernier calliper
深度游标卡尺　depth vernier calliper
高度游标卡尺　heigh vernier calliper
齿厚游标卡尺　gear tooth vernier calliper
带表卡尺　dail calliper
千分尺　micrometer
外径千分尺　outside micrometer
内径千分尺　internal micrometer
公法线千分尺　gear tooth micrometer
塞尺　feeler
直角尺　mechanical square
百分表　dial gage
大量程百分表　long range dial gage
内径百分表　dial bore gage
测微表　micrometer
千分表　dial indicator
量块　gage block
角度块　angular gage block
干涉仪　interferometer
光栅　raster display
系统误差　system error
随机误差　stochastic error
粗大误差　crassitude error
函数误差　function error

Measure is an instrument for pledging part quality.
测量是保证零件质量的一种手段。

第三节 形状和位置公差
Section 3　Geometrical Tolerancing and Location Tolerance

理想要素	ideal feature	倾斜度	angularity
实际要素	real feature/actual factor	定向公差	orientation tolerance
单一要素	singleness essential factor	定位公差	location tolerance
关联要素	associate essential factor	同轴度	concentricity

★ 形状公差　tolerance in form/geometrical tolerance
对称度　symmetry
直线度　straightness
位置度　position
平面度　flatness
跳动公差　run-out tolerance
圆度　roundness/circularity
圆跳动　circular runout
圆柱度　cylindricity
全跳动　total runout
线轮廓度　profile of line
公差原则　tolerancing principle
面轮廓度　profile of plane
独立要求　independent requirement
　　　　　　　　　　　　　　包容要求　envelope requirement
★ 位置公差　tolerance in position/location tolerance
最大实体要求　maximum material requirement
最小实体要求　least material requirement
平行度　parallelism
可逆要求　reciprocity requirement
垂直度　perpendicularity
延伸公差带　projection tolerance zone

Choosing form tolerance and position tolerance is an importance compose content of the precision design.
确定形状公差和位置公差对于保证零件的精度设计是非常重要的。

第四节 表面粗糙度及检测
Section 4　Surface Roughness and Measuring

取样长度　sampling length
评定长度　evalution length
轮廓中线　mean line
★ 轮廓算术平均偏差　Ra
微观不平度十点高度　Rz
轮廓最大高度　Ry
★ 表面粗糙度的测量　surface roughness measuring
表面粗糙度测量仪器　surface roughness measuring instrument
表面粗糙度比较样块　surface roughness comparison specimen
比较法　comparison means
光切法　optical means
干涉法　interferometer means
针描法　needle scanning means

The surface roughness is an index that judges the surface quality of the parts.
表面粗糙度是衡量零件表面质量的一个指标。

第五节 光滑极限量规
Section 5　Smooth Limit Gauge

★ 光滑极限量规　smooth limit gauge
止规　not go gauge
量规　gauge
工作量规　working gauge
塞规　plug gauge
验收量规　reception gauge
环规　ring gauge
校对量规　master gauge
卡规　caliper
花键综合量规　spline gauge
通规　go gauge
位置量规　gauge for measuring position

The smooth limit gauge is a measure without scales.
光滑极限量规是一种没有刻度的量具。

第六节 滚动轴承的公差与配合
Section 6 Tolerance and Fit of Rolling Bearing

内圈　inner ring	轴向载荷　axial load
外圈　outer ring	角接触轴承　angular contact bearing
保持架　cage	深沟球轴承　deep groove ball bearing
滚动体　rolling element	向心轴承　radial bearing
轴承内径　bearing bore diameter	向心滚子轴承　radial roller bearing
轴承外径　bearing outside diameter	圆柱滚子轴承　cylindrical roller bearing
轴承宽度　bearing width	圆锥滚子轴承　tapered roller bearing
滚动轴承　rolling bearing/antifriction bearing	摆动载荷　oscillating load
径向载荷　radial load	

There are four main parts of ball bearings. The inner and outer rings provide the pathway on which the balls move. The retainer or cage maintaining separation of the balls, and a shield to keep dirt from entering or a seal to retain lubrication.

球轴承有四个主要部分。内圈和外圈是为滚珠移动而设置的轨道。保护架是为了将滚珠彼此隔开并保持相对的位置关系而设置的，它还可以防止灰尘进入并保留润滑油。

第七节　尺寸链
Section 7 Dimensional Chain

★尺寸链　dimensional chain	★组成环　component link
尺寸精度　dimensional accuracy	增环　increase link
加工精度　machining accuracy	减环　decrease link
加工误差　machining error	补偿环　compensate link
★环　link	概率法　probability means
★封闭环　closed link	完全互换法　complete interchangeability means

Dimensional chain is the close dimensions that is made up of one another dimension.

由相互连接的尺寸形成的封闭尺寸组称为尺寸链。

第八节　圆锥的公差配合及检验
Section 8 Tolerance and Fit and Test of Taper

圆锥　taper	圆锥配合　cone fit
圆锥角　taper angle	圆锥塞规　plug cone gauge
锥度　taper	圆锥量规　taper gauge
圆锥公差　cone tolerance	

The cone fit is a representative structure commonly used.
圆锥配合是常用的典型结构。

第九节　螺纹公差及检测
Section 9 Tolerance and Test of Threads

紧固螺纹　fasten thread	外螺纹　external thread
传动螺纹　transmission thread	内螺纹　internal thread
紧密螺纹　tightness thread	螺纹牙型　screw thread profile
螺纹　screw thread; thread	粗牙普通螺纹　coarse metric thread
螺纹紧固件　thread fastener	细牙普通螺纹　fine metric thread

梯形螺纹　trapezoidal thread
大径　major diameter
小径　minor diameter
中径　pitch diameter
公称直径　nominal diameter
螺距　pitch
螺纹量规　screw gauge
螺纹塞规　plug cone gauge
螺纹环规　screw ring gauge

The interchangeability and fit character of the common screw thread lie on the pitch diameter.
普通螺纹的互换性和配合性质主要取决于中径。

第十节　键和花键的公差与配合
Section 10　Tolerance and Fit of Key and Splines

花键　spline
★ 键　key
普通平键　flat key
半圆键　round key
钩头楔键　wedge key
导向平键　feather key/dive key

In key coupling, the key width is benchmark, employing shaft-basic system of fits.
在键连接中,键宽是基准,采用基轴制。

第十一节　圆柱齿轮传动公差及检测
Section 11　Tolerance and Test of the Cylindrical Gear Driving

★ 齿轮　gear
圆柱齿轮　cylindrical gear
渐开线圆柱齿轮　involute cylindrical gear
直齿轮　spur gear
斜齿轮　spiral gear
轮齿　gear tooth
节圆　pitch circle
几何偏心　geometric eccentricity
★ 齿圈径向跳动（ΔF_r）radial run-out of gear
★ 公法线长度变动（ΔF_w）variation of base tangent length
齿距累计误差（ΔF_p）total cumulative pitch error
★ 切向综合误差（$\Delta F'_i$）tangential composite error
★ 径向综合误差（$\Delta F''_i$）radial composite error
运动偏心　movement eccentricity
基齿距偏差（Δf_{pb}）base pitch error
齿距偏差（Δf_{pt}）circular pitch individual error
一齿切向综合误差（$\Delta f'_i$）tangential tooth-to-tooth composite error
一齿径向综合误差（$\Delta f''_i$）radial tooth-to-tooth composite error
螺旋线波动误差（$\Delta f_{f\beta}$）helix waveness error
螺旋线总偏差（ΔF_β）total tooth orientation error
接触线误差（ΔF_b）contact line error
齿厚偏差（ΔE_s）deviation of width of teeth
公法线平均长度偏差（ΔE_{wm}）deviation of base tangent mean length

In this section, we mostly understand precision and application on involute cylindrical gears and gear pair.
在本节中我们主要了解了渐开线圆柱齿轮及齿轮副的精度和应用。

第十一章 计算机辅助设计/计算机辅助制造技术
Chapter 11 CAD/CAM Technology

第一节 概述
Section 1 Introduction

计算机辅助制造 computer-aided manufacturing (CAM) | 计算机辅助设计 computer-aided design (CAD)

1 CAD/CAM is a term which means computer-aided design and computer-aided manufacturing. It is the technology concerned with the use of digital computer to perform certain functions in design and production. This technology is moving in the direction of greater integration of design and manufacturing, two activities which have traditionally been treated as distinct and separate functions in a production firm.

CAD/CAM 的意思是计算机辅助设计和计算机辅助制造,就是用数字计算机来完成设计和制造功能。这两种技术正在向着设计和制造的更高层次的集成发展,而这两种技术,以传统的观点看,它们是生产公司中两个截然不同、彼此分离的功能。

2 Computer-aided design (CAD) can be defined as the use of computer systems to assist in the creation, modification, analysis, or optimization of a design.

计算机辅助设计可定义为使用计算机系统来协助一个设计方案的形成、修改、分析及优化。

第二节 CAD/CAM 系统的硬件和软件
Section 2 Hardware and Software of CAD/CAM System

★ 阴极射线管 cathode-ray tube (CRT)
显示处理器 display processor
缓冲器 buffer
虚拟现实 virtual-reality
★ 像素 pixel
位图 bitmap
荧光屏 phosphor-coated screen
加速极 accelerating anode
聚焦系统 focusing system
★ 液晶显示 liquid-crystal display (LCD)
★ 键盘 keyboard
★ 鼠标 mouse
点阵打印机 dot-matrix printer
激光打印机 laser printer
喷墨打印机 ink-jet printer
滚筒式绘图机 rollfeed pen plotter
监视器 monitor
终端 terminal
外围设备 peripheral equipment
数字化仪 digitizer
光笔 light pen
平板笔式绘图机 desktop pen plotter
彩色图形显示适配器 color graphics adapter
视频图形显示适配器 video graphics adapter
扫描仪 scanner
语音识别系统 speech-recognition system
语音系统 voice system

1 Typically, the primary output device in a graphics system is a video monitor. The operation of most video monitors is based on the standard cathode-ray tube (CRT) design, but several other technologies exist and solid-state monitors may eventually predominate.

传统上,图形系统的主要输出设备是视频监视器。大多数视频监视器都是基于标准阴极射线管来设计的,但是也有其他的设计技术存在,固体监视器最终可能会成为主要的监视器。

2 Liquid-crystal displays(LCDs) are commonly used in small systems, such as calculators and portable laptop computers. These nonemissive devices produce a picture by passing polarized light from the

第十一章 计算机辅助设计/计算机辅助制造技术

surroundings or from an internal light source through a liquid-crystal material that can be aligned to either block or transmit the light.

液晶显示器常用在小型系统上，比如：计算器和手提式计算机。这种无辐射设备是通过反射自然光或内部光源发出的光来显示图形的，这种光在通过液晶材料的时候被极化，液晶材料在不同排列方式下可以阻止或允许极化光通过。

第三节 CAD/CAM 系统的开发基础
Section 3 Development Fundamental of CAD/CAM System

★ 二叉树　bintree/binary tree
遍历　searching
遍历树　tree search
子树　subtree
★ 队列　queue
★ 栈　stack
★ 指针　pointer
父节点　parent
子节点　children
表　list table
串　string
根　root

度　level
路径　path
深度　height
排序　sort
软件包　package
工作站　workstation
★ 交钥匙系统　turnkey system
数据库管理系统　data base management system (DBMS)
★ 工程数据库管理系统　engineering data base management system (EDBMS)
面向对象数据库　object-oriented data base

1 The binary search tree gets its tree structure by allowing each node to point to two other nodes: one that precedes it in the list, and one that follows. Unlike a node in a doubly linked list, a node in a binary search tree does not necessarily point to the nodes whose values immediately precede or follows.

二叉树的树状结构是由一些节点组成的，而任一节点都指向其他的两个节点：列表中的前驱节点和后续节点。二叉树中的节点与双向链表中的节点不同，这些节点没必要必须指向值比它大的直接前驱和后续节点。

2 There are two general classifications for graphics software: general programming packages and special-purpose applications packages.

图形软件可分为两大类：通用编程软件包和专用应用软件包。

第四节 计算机图形学
Section 4 Computer Graphics

★ 裁剪操作　clipping operation
裁剪算法　clipping algorithm
裁剪窗口　clipping window
直线裁剪　line clipping
比例　scaling
平行投影　parallel-projection
★ 透视投影　perspective projection
★ 矩阵　matrix
齐次坐标　homogeneous coordinate
镜像　reflection
★ 错切　shear
笛卡尔坐标　Cartesian coordination

世界坐标系　world coordinate
正交坐标系　normalized coordinate
设备坐标系　device coordinate
几何变换　geometric transformation
★ 视区　view port
★ 窗口　window
★ 窗视变换　window-to-view port transformation
窗变换　window transformation
视变换　viewing transformation
所见即所得型　what you see is what you set
用户界面管理系统　user interface management system (UIMS)

图形核心系统　graphical kernel system (GKS)
程序员层次交互图形系统　programmer's hierarchical interactive graphic standard (PHIGS)
计算机图形元文件编码　computer graphics metafile (CGM)
计算机图形接口编码　computer graphics interface (CGI)
初始图形交换规范　initial graphics exchange specification (IGES)
产品模型数据交换标准　standard for the exchange of product model data (STEP)

1 A translation is applied to an object by repositioning it along a straight-line path from one coordinate location to another. We translate a two-dimensional point by adding translation distance: more tx and ty from the original coordinate position (x, y) to a new position (x', y') point.

平移变换就是把一个物体沿直线路径从一个坐标位置移到另一坐标位置。我们可以给一个二维点加上一个平移距离: tx 和 ty,把它从原坐标位置(x,y)移到新的坐标位置(x',y')。

2 The rotation transformation operation $R(\theta)$ is the 3 by 3 matrix in Eq. 5-19 with rotation parameter θ. We get the inverse rotation matrix when θ is replaced with $-\theta$.

在等式 5-19 中,旋转变换运算符 $R(\theta)$ 是一个 3×3 矩阵,其旋转参数是 θ,当用 $-\theta$ 代替 θ 时,就得到了反向旋转矩阵。

第五节　实体建模
Section 5　Object Modeling

★ 实体建模　solid modeling
线框　wireframe
★ 线框建模　wireframe modeling
★ 表面建模　surface modeling
★ 样条　spline
★ 插值　interpolation
★ 拟合　approximation
特征多边形　characteristic polygon
贝塞尔曲线　Bezier curve
★ B 样条　B-spline
扫描　sweep

构造立体几何法　constructive solid geometry (CSG)
混合模式　hybrid model
边界表示法　boundary representation
四叉树　quadtree
八叉树　octree
网格面　mesh
消隐　shading
干涉　interfere
网格　grid

1 Another technique for solid modeling is to combine the volume occupied by overlapping three-dimensional objects using set operations. This modeling method called constructive solid geometry (CSG).

另一个实体建模技术就是一种用集合运算把多个三维实体拼合在一起的方法,这种建模方法叫做构造立体几何法(CSG)。

2 Since the geometric data tables may contain extensive listings of vertices and edges for complex objects, it is important to check the data for consistency and completeness. The more information included in the tables, the easier it is to check for errors.

对于复杂的实体,几何数据表可能包含大量的顶点和边的数据,此时检验数据是否一致和完整是十分重要的。表中的数据越多,越容易检查出错误。

第六节　计算机辅助工程
Section 6　Computer Aided Engineering

★ 计算机辅助工程分析　computer aided engineering (CAE)
★ 有限元　finite element

有限元法　finite element method
★ 自由度　freedom
平衡方程　equation of equilibrium

第十一章　计算机辅助设计/计算机辅助制造技术

总体刚度矩阵　global stiffness matrix
★ 优化　optimization
优化方法　optimal method
设计变量　design variable
设计参数　design parameter
分析函数　analysis function
状态变量　state variable
目标函数　objective function
约束　constraint
边界约束　side constraint
函数约束　functional constraint
仿真　simulation
离散事件　discrete event
连续系统　continuous system
物理仿真　physical simulation

In modern mechanical design it is rare to find a project that does not require some type of finite element analysis (FEA). When not actually required, FEA can usually be utilized to improve the quality a design. The practical advantages of FEA in stress analysis and structural dynamics have made it become the accepted design tool for the last decade.

在现代机构设计中,很少发现不用有限分析(FEA)进行设计的。在不是非常需要使用有限元的场合,有限元分析一般用来提高设计质量。有限元分析在引力分析和结构动态分析中的优越性使得它成为了前十年中被人所接受的设计工具。

第七节　计算机辅助工艺规程设计
Section 7　Computer Aided Process Planning

★ 计算机辅助工艺规程设计　computer aided process planning (CAPP)
★ 检索式CAPP系统　retrieval-type CAPP system
★ 派生式CAPP系统　variant CAPP system
★ 创成式CAPP系统　generative CAPP system
知识库　knowledge base
设计周期　turnaround time
工艺卡　route sheet
知识获取系统　knowledge acquisition system
★ 成组技术　group technology (GT)
逻辑决策　logical decision
编码系统　coding system
零件族　part family
复合样件　composite part
在制品　work-in-process (WIP)
分类　classification
订货到交货的周期　lead time
★ 编码　coding
★ 奥匹兹编码系统　Optiz coding system
柔性编码　flexible coding
刚性编码　rigid coding
★ 决策树　decision tree
★ 决策表　decision table

1 There are two approaches of computer-aided process planning: retrieval-type CAPP system, generative CAPP system.
计算机辅助工艺规程设计有两种方法:检索式CAPP系统和创成式CAPP系统。

2 The generative CAPP system synthesizes the design of the optimum process sequence, based on an analysis of part geometry, material, and other factors which would influence manufacturing decisions.
创成式CAPP系统融进了最优工艺流程设计,它是基于零件几何形状、材料和其他可能影响生产决策因素分析基础上形成的一种计算机辅助工艺规程。

第八节　CAD/CAM集成和计算机集成制造系统
Section 8　CAD/CAM Integration and CIMS

美国机械工程师学会　American Society of Mechanical Engineers (ASME)
美国国家标准协会　American National Standards Institute (ANSI)
数字控制　numerical control (NC)
★ 自动编程工具　automatically programmed tool (APT)
计算机辅助夹具设计　computer aided fixture de-

sign (CAFD)
- ★ 计算机辅助质量控制　computer aided quality control (CAQC)
- ★ 计算机辅助检测　computer aided inspection

制造单元　manufacturing cell (MC)
- ★ 柔性制造系统　flexible manufacturing system (FMS)

计算机集成制造　computer integrated manufacturing (CIM)
- ★ 计算机集成制造系统　computer integrated manufacturing system (CIMS)

制造自动化协议　manufacturing automation protocol (MAP)

技术和办公协议　technical and office protocol (TOP)

物料需求计划　material require planning (MRP)

制造资源计划　manufacturing resource planning (MRPII)
- ★ 准时生产　just in time
- ★ 专家系统　expert system
- ★ 精良生产　lean production
- ★ 虚拟制造　virtual manufacturing
- ★ 敏捷制造　agile manufacturing
- ★ 并行工程　concurrent engineering
- ★ 逆向工程　reverse engineering
- ★ 快速原形　rapid prototyping

产品数据管理　product data management (PDM)

机器人　robot

机器人学　robotics
- ★ 机电一体化　mechatronics

1 Computer-integrated manufacturing or CIM is used to describe the most modern approach of manufacturing. Although CIM encompasses many other advanced manufacturing technologies such as computer numerical control (CNC), computer-aided design/computer-aided manufacturing (CAD/CAM), robotics, and just-in-time delivery (JIT), it is more than a new technology or a new concept.

计算机集成制造(CIM)用于描述最现代的制造方法。虽然 CIM 包含了很多其他的先进制造技术，如：计算机数字控制(CNC)、计算机辅助设计／计算机辅助制造(CAD/CAM)、机器人、准时生产(JIT)等，但它却不仅仅是一门新技术和新概念。

2 The CIM wheel includes five distinct components: general business management, product and process definition, manufacturing planning and control, factory automation, information resource management.

CIM 轮盘图包括五个不同的组成部分：通用商务管理、产品和工艺定义、制造计划和控制、工厂自动化、信息资源管理。

第十二章 非传统加工/特种加工
Chapter 12 Non-conventional Machining/Non-traditional Machining

第一节 绪论
Section 1 Introduction

超精密　ultraprecise
电能　electrical energy
混杂的　miscellaneous
电加工工艺　electrical-machining processes
难加工材料　difficult-to-machine material
复杂几何形状　complex geometrical shape
脆性零件　fragile part
脆性材料　brittle material
硬脆材料　hard-brittle material
刀具磨损　tool wear
微机械加工　micromachining

Several of them have been developed largely because of the need to shape new high strength and temperature resistant alloys that are not easily worked by the older processes.
由于出现了用传统加工工艺不易加工的新型高强度和耐高温合金成型的需求，因而几种新型加工工艺得到了很大发展。

第二节 电火花加工
Section 2 Electro-discharge machining

电火花加工　electro-discharge machining
放电　discharge
火花　spark
腐蚀　erosion
腐蚀的　erosive
退化、变质　degenerate
型腔　cavity
退化(变质)成　degenerate into
碎片　debris
华氏温度　fahrenheit
冶金的　metallurgical
电流　current
间隙　gap
气化　vaporize
成型电极　shaped electrode
石墨　graphite
铜　copper
不导电液体　nonconductive liquid
相反形状的　reverse shape
阴极　cathode
阳极　anode
直流(电)　DC current

It must be noted that if a spark is permitted to last too long, it will degenerated into a stationary arc which is not suitable for machining.
必须注意的是，如果电火花持续的时间过长，它将会退化成稳定的电弧，这种电弧不适于加工。

第三节 电火花线切割
Section 3 Electrical Discharge Wire Cutting

电火花线切割　electrical discharge wire cutting
内应力　internal stress
完整的　integral
派生、衍生　variant

The wire, which is constantly fed from a spool, is held between upper and lower diamond guides. The guides, usually CNC-controlled, move in the x-y plane. On most machines, the upper guide can also move independently in the z-u-v axis, giving rise to the ability to cut tapered and transitioning shapes (circle on the bottom square at the top for example). The upper guide can control axis movements in x-y-u-v-i-j-k-l. This allows the wire-cut EDM to be programmed to cut very intricate and delicate shapes.

电火花线由上下两个金刚石导轮夹持着,从线轴连续不断地送入加工区。导轮一般由CNC控制,在x-y平面内移动。在大多数机床上,上导轮可以单独沿z-u-v轴移动,这样的机床具有加工锥形和过渡形状(比如,上圆下方的形状)能力。导轮可以沿x-y-u-v-i-j-k-l轴移动,这使得电火花线切割可以加工非常复杂的形状。

第四节 电化学加工
Section 4 Electrochemical Machining

安培 amperage	除(去)镀(层)工艺 deplating process
溶解 dissolve	电化学磨削 electrochemical grinding
导电的 electrically conductive	电化学珩磨 electrochemical honing
电化学加工 electrochemical machining	电化学车削 electrochemical turning
电解 electrolysis	电化学去毛刺加工 electrochemical deburring
电解的 electrolytic	毛刺 burrs
电解液、电解质 electrolyte	表面精度 surface finish
磨料、磨料的 abrasive	化学溶液 chemical solution

The electrolyte is fed between the wheel and work surface in the direction of movement of the wheel periphery so that it is carried paste the work surface by the action of the wheel rotation.

在砂轮和工件表面之间,电解液沿着砂轮边缘的运动方向注入,通过砂轮旋转作用,电解液流过工件表面。

第五节 激光束加工
Section 5 Laser Beam Machining

激光束加工 laser beam machining (LBM)	基态 ground state
光子 electron	聚焦透镜 focusing lens
相干性 coherency	陶瓷 ceramic
单色的 monochromatic	硅 silicon
单色相干光 monochromatic coherent light	半导体晶片 semiconductor wafer
发散 divergence	晶金刚石 polycrystalline diamond
激发态 excited state	

The basic principle utilized in LBM is that under proper conditions light energy of a particular frequency is used to stimulate the electrons in an atom to emit additional light with exactly the same characteristics of the original light source.

激光加工的基本原理是在一定条件下采用特定频率的光激发原子中电子,使其发射出特征与原光源完全相同的光。

第六节 电子束和离子束加工
Section 6 Electron-beam and Ion-beam Machining

电子束 electron-beam	融化和气化 melt and vaporize
真空腔(室) vacuum chamber	退火 annealing
磁铁 magnet	焊接 welding
使……偏移 deflect	离子束 ion beam
电子(束)流 stream of electron	穿透 penetrate
热能 thermal energy	粒子 particle

第十二章　非传统加工/特种加工

轰击　bombard
原子喷射　atomic sandblasting
光敏的　photo sensitive

微波电路　microwave circuit
定制薄膜电路　custom film circuit

In electron-beam machining (EBM), electrons are accelerated to a velocity nearly three-fourths that of light ($\approx$ 200 000 km/sec). The process is performed in a vacuum chamber to reduce the scattering of electrons by gas molecules in the atmosphere. The electron beam is aimed using magnets to deflect the stream of electrons and is focused using an electromagnetic lens.

在电子束加工中,电子被加速到接近光速的四分之三($\approx$ 200 000 千米/秒)。为了消除空气中的分子对电子的散射作用,整个过程在真空室内完成。电子束靠磁场偏转,靠电磁透镜聚焦。

第七节　超声加工
Section 7　Ultrasonic Machining

超声波的、超音速的　ultrasonic
磨料悬浮液　abrasive slurry
振动　vibration
淬火钢　hardened steel
硬质合金　carbide
红宝石　ruby

石英　quartz
金刚石　diamond
玻璃　glass
超声清洗　ultrasonic cleaning
超声变幅杆　ultrasonic horn

Ultrasonic machining is a mechanical metal removal process for brittle materials which uses high frequency oscillations of a shaped tool using an abrasive slurry. The term ultrasonic refers to the frequency range above the audible range and is above 16kHz.

超声加工是依靠成形工具在磨料悬浮液中高频振动来从脆性材料上去除材料的加工方法。"超声"一词是指振动频率超出了人的听觉范围,即16kHz。

第八节　快速成型技术
Section 8　Rapid Prototyping Technique

光敏树脂液相固化成形　stereo lithography
选择性激光粉末烧结成形　selected laser sintering
薄片分层叠加成形　laminated object manufacturer
熔丝堆积成形　fused deposition modeling
烧结　sintering
分层　slice
截面　cross-sectional plane
一层一层地　layer by layer

固化　solidified
光的波长　wavelength of light
光敏树脂　photopolymer
热塑性塑料　thermoplastic
三维CAD模型　3D CAD model
粉末　powder
加热喷头　heated extruding head
丙烯酸树脂　acrylate
光敏的　photo-curable
光子　photon

Stereo lithography apparatus (SLA) was invented by Charle Hull of 3D Systems Inc.. It is the first commercially available rapid prototype and is considered as the most widely used prototyping machine. The material used is liquid photo-curable resin, acrylate. Under the initiation of photons, small molecules (monomers) are polymerized into large molecules. Based on this principle, the part is built in a vat of liquid resin.

光固化成型机由3D System公司的Charle Hull发明。它是世界上第一台使用最广泛的商业快速成型机。使用的材料为丙烯酸光敏树脂,在光子的引发下,小分子(单体)聚合成大分子,原型零件就是根据这一原理从树脂槽里制作出来的。

第九节 特种加工方法
Section 9 Non-conventional Machining Method

化学加工　chemical machining
化学腐蚀　chemical etching
化学铣削　chemical milling
光化学加工　optical chemical machining
化学抛光　chemical polish
等离子体　plasma
等离子流　plasma-jet
等离子体加工　plasma arc machining
磨料流加工　abrasive flow machining
水射流切割　water jet cutting
磁性磨料研磨加工　magnetic abrasive machining
磁性磨料电解研磨加工　magnetic abrasive electrochemical machining

1 The water jet acts like a saw and cuts a narrow groove in the material. A pressure level of about 400MPa (60 ksi) is generally used for efficient operation, although pressure as high as 1 400MPa (200 ksi) can be generated. Jet-nozzle diameters range between 0.05 mm and 1 mm (0.002 in. and 0.04 in.).

水射流像一个锯一样在材料上切出窄槽。虽然水射流使用的压力可达1 400MPa(200ksi)，但为了得到一定的操作效率，压力一般在400MPa(60ksi)。喷嘴直径在0.05毫米到1毫米(0.002英寸和0.04英寸)。

2 The elimination of the mechanical stage also overcomes one difficulty inherent in conventional machining, namely the increase in tool forces and tool wear encountered when machining the harder metals and alloys used in modern engineering practice.

在加工阶段要消除电能转换为机械能效率低的现象还需克服传统加工工艺中存在的一个固有缺点，即当加工现代工程实际中使用高硬度金属和合金时碰到的切削力大和刀具磨损增加的问题。

3 Ultrasonic machining, electron-beam machining, plasma-jet machining, and laser machining are all examples of these new processes.

超声波加工、电子束加工、等离子加工和激光加工都是这些新型加工工艺的范例。

第十三章 塑料成型与模具制造
Chapter 13 Plastic Deformation and Mould Manufacturing

第一节 注射模具
Section 1 Injection Molds

1.1 The Injection Molding 注射成型

Injection molding is principally used for the production of the thermoplastic parts, although some progress has been made in developing a method for injection molding some thermosetting materials. The problem of injecting a melted plastic into a mold cavity from a reservoir of melted material has been extremely difficult to solve for thermosetting plastics which cure and harden under some conditions within a few minutes. The principle of injection molding is a quite similar to that of die-casting. The process consists of feeding a plastic compound in powered or granular form from a hopper through metering and melting stages and then injecting it into a mold. After a brief cooling period, The mold is opened and solidified part ejected.

尽管成型某些热固性材料的方法取得了一定的进步,但注射成型主要用于生产热塑性塑件。这主要是因为热固性塑料熔体在很短的时间内就会固化和硬化,在从料斗向模具型腔注入热固性塑料熔体的过程中,也会出现这种现象,这个问题一直非常难解决。注射成型原理和铸造十分相似。注射成型工艺过程包括:首先把料斗中的粉状或粒状的塑料依次输送到计量区和熔化区,然后注射到模具的型腔中,经过短时冷却后开模,推出成型塑件。

A typical injection molding cycle or sequence consists of five phases (see Fig. 13-1):
(1) Injection or mold filling;
(2) Packing or compression;
(3) Holding;
(4) Cooling;
(5) Part ejection.

一个典型的注射成型周期或顺序由下列五个阶段组成(见图 13-1):
(1) 注射或模具填充;
(2) 压实或压缩;
(3) 保压;
(4) 冷却;
(5) 分模并顶出。

1.2 Injection Molds 注射模具

Molds for injection molding are as varied in design, degree of complexity, and size as are the parts produced from them. The functions of a mold for thermoplastics are basically to impart the desired shape to the plasticized polymer and then to cool the molded part.

用于注射成型的模具结构和复杂性皆不相同,其尺寸大小也因它所成型的零件的不同而不同。从根本上讲热塑性模具的功能就是赋予增强塑料所要求的形状,然后冷却而形成成型零件。

A mold is made up of two sets of components:(1) the cavities and cores, (2) the base in which the cavities and cores are mounted. The size and weight of the molded parts limit the number of cavities in the mold and also determine the equipment capacity required. From consideration of the molding process,

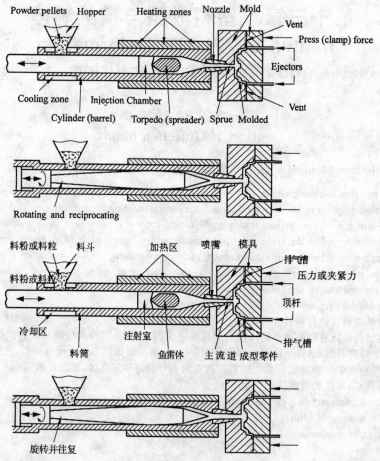

Fig. 13-1 Injection molding process
图 13-1 注射成型工艺

a mold has to be designed to safely absorb the forces of clamping, injection, and ejection. Also, the design of the gates and runners must allow for efficient flow and uniform filling of the mold cavities.

模具由两种零件组成:(1) 型腔和型芯;(2) 型腔和型芯赖以安装的基础。成型零件的尺寸和质量限制了模具中的型腔数目,同时也确定了设备的性能规格。出于成型工艺的考虑,一副模具的设计结构必须能够安全吸收夹紧力、注射力以及顶出力等。与此同时,浇口和流道必须允许高效流动并均匀填充模具型腔。

Fig. 1-2 illustrates the parts in a typical injection mold. The mold basically consists of two parts: a stationary half (cavity plate), on the side where molten polymer is injected, and a moving half (core plate) on the closing or ejector side of the injection molding equipment. The separating line between the two mold halves is called the parting line. The injected material is transferred through a central feed channel, called the sprue. The sprue is located on the sprue bushing and is tapered to facilitate release of the sprue material from the mold during mold opening. In multicavity molds, the sprue feeds the polymer melt to a runner system, which leads into each mold cavity through a gate.

图 13-2 说明了一副典型注射模具的零件组成。这副模具基本由两部分组成:静止部分(型腔板),在这一侧注射熔融的塑料,以及运动部分(型芯板),这是注射成型设备的闭合侧或顶出侧。两半模具的分

第十三章 塑料成型与模具制造

离线称为分型线。注射材料通过中心流道进行传递,称之为主流道。主流道定位于浇口衬套中并具有一定的锥度,以利于模具打开时能够释放来自模具的流道材料。在多腔模具中,主流道把熔融塑料输送到流道系统中,流道系统通过浇口把熔融塑料引入到每个型腔中。

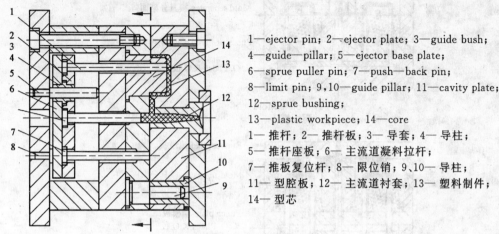

1—ejector pin; 2—ejector plate; 3—guide bush;
4—guide—pillar; 5—ejector base plate;
6—sprue puller pin; 7—push—back pin;
8—limit pin; 9,10—guide pillar; 11—cavity plate;
12—sprue bushing;
13—plastic workpiece; 14—core
1—推杆;2—推杆板;3—导套;4—导柱
5—推杆座板;6—主流道凝料拉杆;
7—推板复位杆;8—限位销;9、10—导柱;
11—型腔板;12—主流道衬套;13—塑料制件;
14—型芯

Fig. 13-2 Injection mold
图 13-2 注射模具

    The core plate holds the main core. The purpose of the main core is to establish the inside configuration of the part. The core plate has a backup or support plate. The support plate in turn is supported by pillars against the U-shaped structure known as the ejector housing, which consists of the rear clamping plate and spacer blocks. This U-shaped structure, which is bolted to the core plate, provides the space for the ejection stroke also known as the stripper stroke. During solidification the part shrinks around the main core so that when the mold opens, part and sprue are carried along with the moving mold half. Subsequently, the central ejector is activated, causing the ejector plates to move forward so that the ejector pins can push the part off the core. Both mold halves are provided with cooling channels through which cooled water is circulated to absorb the heat delivered to the mold by the hot thermoplastic polymer melt. The mold cavities also incorporate fine vents (0.02 to 0.08 mm by 5 mm) to ensure that no air is trapped during filling.

    型芯板夹持主型芯。主型芯的作用是建立零件的内部结构。型芯板具有一个背板或支撑板。支撑板反过来受到靠在 U 形结构上的支柱所支撑,该 U 形结构就是已知的支架,它由后夹持板和垫块组成。U 形结构使用螺栓连接到型芯板上,提供推出行程的空间,也称之为卸料行程的间隔。在固化阶段,零件收缩包覆在主型芯上,于是模具打开时,零件和浇口凝料随动模部分一起带出。接着,中心推出器开始作用,引起顶板向前移动,从而推杆使零件脱离型芯。模具的两个部分都提供有冷却水道,冷却水循环带走热塑性熔融塑料带给模具的热量。模型型腔还引入了细小的排气口(0.02~0.08 mm 深,5 mm 宽),以确保在填充过程中没有空气因入到模具中。

    There are six basic types of injection molds in use today. They are: (1) two-plate mold; (2) three-plate mold, (3) hot-runner mold; (4) insulated hot-runner mold; (5) hot-manifold mold; and (6) stacked mold. Fig. 13-3 and Fig. 13-4 illustrate these six basic types of injection molds.

    现今使用中的注射模具具有六种基本类型,它们是:(1)两板模;(2)三板模;(3)热流道模;(4)绝热热流道模;(5)热歧管模和(6)层叠模。图 13-3 和图 13-4 说明了六种基本类型的注射模具。

Injection Mold Dies   注射模具
See Fig. 13-3 for the other three types.
关于其他三种类型见图 13-3。

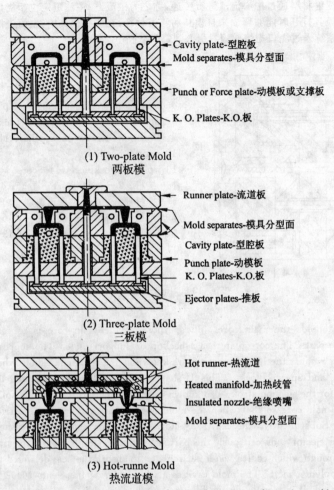

Fig. 13-3 This illustrates three of the six basic types of injection mold
图 13-3　六种基本类型注射模具的三种图解

See Fig. 13-4 for the other three types.
关于其他类型的注射模具见图 13-4。

(1) Two-Plate Mold　两板模

A two-plate mold consists of two plates with the cavity and cores mounted in either plate. The plates are fastened to the press platens. The moving half of the mold usually contains the ejector mechanism and the runner system. All basic designs for injection molds have this concept. A two-plate mold is the most logical type of tool to use for parts that require large gates.

两板模具是由分别装入型腔和型芯的两块板组成。这两块板安装到压力板上。运动部分通常包括推出机构和流道系统。所有注射模具的基本结构都具有该设计概念。对于要求大浇口的零件成型，两板模则是最合乎逻辑的成型工具。

(2) Three-Plate Mold　三板模

This type of mold is made up of three plates：(1) the stationary or runner plate is attached to the stationary platen，and usually contains the sprue and half of the runner；(2) the middle plate or cavity

第十三章 塑料成型与模具制造

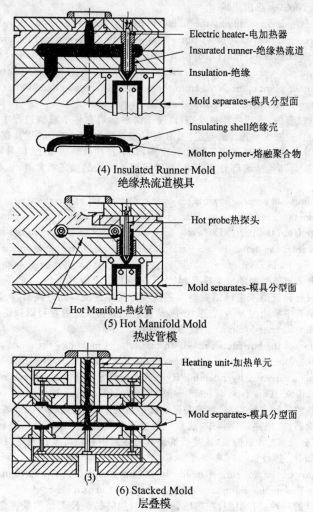

Fig. 13-4 This illustrates three of the six basic types of injection molding dies
图 13-4 六种基本类型注射模具的三种

plate, which contains half of the runner and gate, is allowed to float when the mold is open; and (3) the movable plate or force plate contains the molded part and the ejector system for the removal of the molded part. When the press starts to open, the middle plate and the movable plate move together, thus releasing the sprue and runner system and degating the molded part. This type of mold design makes it possible to segregate the runner system and the part when the mold opens. The die design makes it possible to use center-pin-point gating.

该类型的模具由三块板组成:(1) 连接在固定安装上的定模板或流道板,通常包含主流道和一半的分流道;(2) 中间板或型腔板,通常包含一半的分流道和浇口,当模具打开时允许其浮动;(3) 移动板或支撑板包含成型零件以及用于成型零件取出的推出系统。当压力机开始打开,中间板和移动板一起移动,于是分离主流道和流道系统,并切断成型零件浇口。该类型的模具结构使开模时分离流道系统和零件成为可能。该种模具结构适用于中心点浇口成型。

(3) Hot-Runner Mold　热流道模

In this process of injection molding, the runners are kept hot in order to keep the molten plastic in a fluid state at all times. In effect this is a "runnerless" molding process and is sometimes the same. In runnerless molds, the runner is contained in a plate of its own. Hot runner molds are similar to three-plate injection molds, except that the runner section of the mold is not opened during the molding cycle. The heated runner plate is insulated from the rest of the cooled mold. Other than the heated plate for the runner, the remainder of the mold is a standard two-plate die.

在该种注射成型过程中,流道保持热状态,以便于熔融塑料在整个成型过程中都维持在流体状态。就该种效果而言,这是一种"无流道"成型工艺。有些时候这两种叫法是等同的。在无流道模具中,流道包含在各自的板中。热流道模具类似于三板注射模具,只不过在模塑成型中,模具的流道截面不是打开的。加热的流道板和模具其他冷的部分是绝缘的。除了用于流道的加热板之外,该种模具的其余部分则是标准的两板模具。

Runnerless molding has several advantages over conventional sprue runner-type molding. There are no molded side products (gates, runners, or sprues) to be disposed of or reused, and there is no separating of the gate from the part. The cycle time is only as long as is required for the molded part to be cooled and ejected from the mold. In this system, a uniform melt temperature can be attained from the injection cylinder to the mold cavities.

和传统的主流道分流道类型的成型模具相比,无流道模具具有几个优点。没有要进行处理或回收的成型侧的废品(浇口、分流流道或主流道凝料),并且不需要零件浇口的分离机构。成型周期的时间仅是成型零件冷却和从模具中推出的时间。在系统中,从注射缸到模具型腔都可以获得均匀熔融温度。

(4) Insulated Hot-Runner Mold　绝热热流道模

This is a variation of the hot-runner mold. In this type of molding, the outer surface of the material in the runner acts as an insulator for the molten material to pass through. In the insulated mold, the molding material remains molten by retaining its own heat. Sometimes a torpedo and a hot probe are added for more flexibility. This type of mold is ideal for multicavity center-gated parts.

这是热流道模具的一种变种类型。在该类型的成型过程中,分流道材料的外表面的作用类似于熔融材料流经通道的绝缘器。在绝热的模具中,成型材料由于自身的热量而维持熔融状态。有时为了更加灵活而添加了鱼雷体(分流锥)和热探头。该类型的模具是多腔中心浇口零件的理想成型设备。

(5) Hot-Manifold　热歧管

This is a variation of the hot-runner mold. In the hot-manifold die, the runner and not the runner plate is heated. This is done by using an electric-cartridge-insert probe.

这是热流道模具的一种变种类型。在热歧管模具中,加热的是流道而不是流道板。这是通过使用一种电动盒式潜入探头来实现的。

(6) Stacked Mold　层叠模具

The stacked injection mold is just what the name implies. A multiple two-plate mold is placed one on top of the other. This construction can also be used with three-plate molds and hot-runner molds. A stacked two-mold construction doubles the output from a single press and reduces the clamping pressure required to one half, as compared to a mold of the same number of cavities in a two-plate mold. This method is sometimes called "two-level molding".

层叠模具就像它名字所暗示的那样。多个两板模具彼此重叠放在一起。这种结构也随三板模具和热流道模具使用。和相同型腔数目的两板模具相比,层叠的两模结构对于一个压制过程而成倍的输出,并且降低了每一半模具所需要的夹持力。这种方法有时也称之为"两级模塑成型"。

1.3　Injection-molding Machine　注射成型机

Several methods are used to force or inject the melted plastic into the mold. The most commonly used system in the larger machines is the in-line reciprocating screw, as shown in Fig. 1-5. The screw

第十三章　塑料成型与模具制造

acts as a combination injection and plasticizing unit. As the plastic is fed to the rotating screw, it passed through three zones as shown: feed, compression, and metering. After the freed zone, the screw-flight depth is gradually reduced, forced the plastic to compress. The work is converted to heat by shearing the plastic, making it a semifluid mass. In the metering zone, additional heat is applied by conduction from the barrel surface. As the chamber in front of the screw becomes filled, it forced the screw back, tripping a limit switch that activates a hydraulic cylinder that forces the screw forward and injects the fluid plastic into the closed mold. An antiflowback valve prevents plastic under pressure from escaping back into the screw flights.

熔融塑料注入模具通常有几种方式。在大型注塑机上常采用往复螺杆式注入的方式，如图 1-5 所示。螺杆同时具有注射和塑化的功能。树脂原料进入旋转的螺杆时，要经过图示的三个区域：喂入区、压实区和计量区。经过喂入区后，为压实树脂原料，螺杆螺旋部分的深入逐渐降低，同时传递树脂原料间因剪切作用而产生的热量，使原料呈半流动状态。在计量区，螺缸表面的加热装置对熔体进一步加热。当熔体充满螺杆前部区域时，螺杆在熔体压力下后退，触动限位开关使液压缸工作，在液压力的作用下推动螺杆向前运动，将熔融塑料注射到闭合的模具型腔中。防倒流阀能阻止受压熔体倒流进入螺杆的螺旋区。

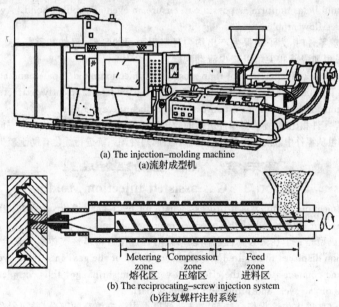

(a) The injection-molding machine
(a)流射成型机

(b) The reciprocating-screw injection system
(b)往复螺杆注射系统

Fig 13-5　The injection-molding system
图 13-5　注射成型系统

第二节　重叠成型
Section 2　Overmolding

1 Overmolding is an injection molding process where two materials are molded together. Types of overmolding include two shot sequential overmolding, multi-shot injection molding or insert overmolding.

重叠注塑是一种注射成型工艺，在该工艺中，两种材料将浇铸在一起。重叠注塑的类型包括二次顺序重叠注塑、多射注射成型或者镶件重叠注塑。

2 Multi-shot injection molding injects multiple materials into the cavity during the same molding cycle. Insert overmolding uses a pre-molded insert placed into the mold before injecting the second material.

Two shot sequential overmolding is where the molding machine injects the first material into a closed cavity, and then moves the mold or cores to create a second cavity, using the first component as an insert for the second shot using a different material.

多射注射成型会在同一个成型周期中将多种材料注射到型腔中。镶件重叠注塑会在注射第二种材料之前使用放在模具中的预成型镶件。在二次顺序重叠注塑过程中,注塑机会先将第一种材料注射到一个封闭的型腔中,然后通过移动模具或型芯来创建第二个型腔,同时将第一个组成作为第二次注射另一种材料的镶件。

3 Materials are usually chosen specifically to bond together, using the heat from the injection of the second material to form that bond. This avoids the use of adhesives or assembly of the completed part. It can result in a robust multi-material part with a high quality finish.

通常要对材料进行特殊的选择以使它们能够通过热量(注射第二种材料时所产生的热量)结合在一起。这样可以避免使用黏合剂或装配已加工成型的零件。这样可以生产出坚固耐用的高光洁度多材料零件。

4 When designing an overmolded part, wall thicknesses of both the insert and the overmolded component should be as uniform as possible to ensure an even and robust bond. Avoid ribs and sharp corners to reduce flow problems.

在设计重叠注塑零件时,为了确保生产出均匀且坚固耐用的零件,镶件和重叠注塑组件的壁厚应尽量均匀。避免采用加强筋和尖角以减少流动问题。

5 Overmolded parts take longer to cool than single shot injection molded part, and cooling systems are less effective. The insert acts as an insulator and heat is less efficiently extracted from the part. However, optimizing the cooling system can help reduce the cycle time.

与单射注射成型零件相比,重叠注塑零件所需的冷却时间更长并且冷却系统的效率更低。镶件起着绝缘体的作用,使得热量从零件中散发出去的效率更低。然而,对冷却系统进行优化有助于减少周期时间。

第三节 气体辅助注射成型
Section 3 Gas-assisted Injection Molding

1 Gas-Assisted Injection Molding is a process where an inert gas is introduced at pressure, into the polymer melt stream at the end of the polymer injection phase.

气体辅助注射成型是一个在聚合物注射阶段结束时通过压力将惰性气体引入聚合物熔体流的过程。

2 The gas injection displaces the molten polymer core ahead of the gas, into the as yet unfilled sections of the mold, and compensates for the effects of volumetric shrinkage, thus completing the filling and packing phases of the cycle and producing a hollow part.

气体注射可将气体前方的熔化聚合物型芯推动到模具中尚未填充的部分,并可补偿体积收缩率的影响,从而完成周期中的填充和保压阶段,生产出空心零件。

3 Traditionally, injection molded components have been designed with a relatively constant wall thickness throughout the component. This design guideline helps to avoid major flaws or defects such as sink marks and warpage. However, apart from the simplest of parts, it is impossible to design a component where all sections are of identical thickness. These variations in wall thickness result in different sections of the part packing differently, which in turn means that there will be differentials in shrinkage throughout the molding and that subsequently distortion and sinkage can often occur in these situations.

过去,注射成型组件的设计是,整个组件中的壁厚保持相对一致。此设计准则有助于避免产生重大瑕疵或缺陷,如缩痕和翘曲。但是,除非零件及其简单,否则不可能设计出所有部分厚度均相同的组件。此类壁厚差异将导致零件不同部分的保压不同,而保压不同反过来又意味着在整个成型过程中收缩率将会不同,从而又会导致常发生在此类情况下的变形和下沉。

第十三章　塑料成型与模具制造

4 By coring out the melt center, gas injection molding enables the packing force (which compensates for differential shrinkage) to be transmitted directly to those areas of the molding which require attention. This dramatically reduces differentials in shrinkage and thus the sinkage. In addition, the internal stresses are kept to a minimum, considerably reducing any distortion that may otherwise have taken place.

通过移除熔体中心，气体注射成型会将保压作用力（补偿收缩率的不同）直接传递到那些需要注意的成型区域。这会显著降低收缩率的差异，从而使下沉几率也随之降低。此外，内部应力将保持在最小值，否则便无法显著减少形变。

5 Maximum clamp pressures are normally required during the packing phase of a molding cycle. This is due to the force which has to be exerted at the polymer gate in order to pack melt into the extremities of the mold cavity in an effort to compensate for the volumetric shrinkage of the solidifying melt. In comparison to compact injection molding, gas injection molding typically has considerably shorter distance over which the solidifying melt is required to be packed because of the gas core. This means that proportionally lower packing pressures are required to achieve the same results and in turn, lower machine clamp forces are required.

在成型周期中的保压阶段，锁模压力通常需要达到最大值。这是因为在聚合物浇口处施加的力可将熔体推入模具型腔的各个末端，以试图补偿固化熔体的体积收缩率。与紧凑的注射成型相比，气体注射成型通常会由于气体型芯的存在而显著缩短固化的熔体需要推进的距离。这意味着达到相同效果所需的保压压力将适当降低，而反过来，所需的机器锁模力也会随之降低。

6 Gas injection allows cost effective production of components with:
- Thick section geometry;
- No sink marks;
- Minimal internal stresses;
- Reduced warpage;
- Low clamp pressures.

气体注射可通过以下途径生产出经济实惠的组件：
- 几何截面厚；
- 无缩痕；
- 内部应力降至最低；
- 减少翘曲；
- 锁模压力较低。

第四节　共注射成型
Section 4　Co-injection Molding

1 Co-injection molding involves the injection of two different materials into the one mold.
共-注成型涉及将两种不同的材料注射到一个模具中。

2 Co-injection molding overview
共-注成型概述

3 Co-injection molding involves injection of two dissimilar materials. Because of this, co-injection has some special advantages, as well as some potential molding problems. The Co-injection analysis helps you overcome the potential problems and leverage the advantages, by helping to optimize process control strategies and enhance part quality.

共-注成型涉及两种不同材料的注射。因此，共-注不仅具有一些特殊的优点，还存在一些潜在的成型问题。通过帮助优化工艺控制策略和提高零件质量，共-注分析有助于克服潜在的问题并充分利用优点。

第五节 注射压缩成型工艺
Section 5　The Injection-compression Molding Process

1 The Injection-compression molding process is an extension of conventional injection molding. After a pre-set amount of plastic melt is fed into an open cavity, it is compressed. The primary advantage of this process is the ability to produce dimensionally stable, relatively stress-free parts, at a low clamp force. Injection-compression molding is sometimes called coining, stamping, compressive-fill, or hybrid molding.

注射-压缩成型工艺是传统注射成型的扩展。预设数量的塑料熔体被浇注到打开的型腔中之后,压缩过程即可开始。该工艺的主要优势是能够以较低的锁模力生产出尺寸稳定、相对无应力的零件。注射-压缩成型有时也称为印压、冲压、压缩性填充或混合成型。

2 Injection-compression simulates the following special characteristics of the injection-compression molding process:

注射-压缩分析可模拟以下注射-压缩成型工艺的特性:

Injection phase

During this stage, the mold cavity thickness is designed to be larger than the target part thickness, in order to allow plastic to flow easily to the extremities of the cavity. Because the plastic flows easily, it can do so under relatively low pressure and stress.

注射阶段

在此阶段,模具型腔厚度被设计成大于目标零件的厚度,其目的是为了使塑料更容易流到型腔的各个末端。因为如果塑料容易流动,那么即使在相对较低的压力和应力下,流动性也不会受到影响。

Compression phase

During or after filling, a compressive force reduces the mold cavity thickness, forcing the resin into the unfilled portions of the cavity. This produces a more uniform packing pressure across the cavity. These results in more homogeneous physical properties and less molded-in stresses compared to conventional injection molding.

压缩阶段

填充时或填充后,压缩作用力可使模具型腔厚度减小,从而使树脂得以进入型腔中尚未填充的部分。这会在整个型腔中产生更加均匀的保压压力。因而同传统注射成型相比,物理属性更均匀,模中应力更小。

Advantages and applications

Injection-compression is advantageous for production of precision parts that require low residual stresses, such as optical discs, and high-precision moldings. Conventional injection molding may not be able to meet product design requirements for these parts because thermoplastics are inherently difficult to process due to their PVT characteristics and high viscosity.

优点与应用

注射-压缩有利于生产需要较低残余应力的精密零件,如光碟和高精度成型物。传统注射成型可能无法满足此类零件在产品设计方面的需求,因为热塑性塑料本身具有 PVT 特性且黏度较高,很难加工。

第六节 反应成型工艺
Section 6　Reactive Molding Process

1 Reactive Molding process, also called thermoset molding process, use thermoset materials.

反应成型工艺(也称为热固性成型工艺)使用热固性材料。

Thermosets, unlike thermoplastics, are characterized by the following:
- A chemical reaction during the molding process

- Cross-linked polymer structure
- Simultaneous polymerization and shaping during the molding process.

热固性与热塑性在以下几个方面有所不同：
- 成型过程期间的化学反应
- 交链聚合物结构
- 成型过程期间同时聚合与塑形。

2 Process

The Reactive Molding processes include the following:
- Reaction Injection Molding (RIM)
- Structural Reaction Injection Molding (SRIM)
- Resin Transfer Molding (RTM)
- Multiple-barrel reactive molding (RIM-MBI)
- Thermoset injection molding
- Rubber injection molding
- Microchip Encapsulation
- Underfill Encapsulation

工艺

反应成型工艺包括以下类别：
- 反应注射成型(RIM)
- 结构化反应注射成型(SRIM)
- 树脂传送成型(RTM)
- 多料筒反应成型(RIM-MBI)
- 热固性塑料注射成型
- 橡胶注射成型
- 微芯片封装
- 底层覆晶封装

3 Advantages

The Reactive Molding analysis offers the following advantages:
- Thermosets' cross-linked polymer structure generally imparts improved mechanical properties and greater heat and environmental resistance.
- Thermosets' typically low viscosity permits large and complex parts to be molded with relatively lower pressure and clamp force than required for thermoplastics molding.
- Thermosets can be used in composite processes. For example, RTM and SRIM processes, which use a preform made of long fibers, offer a way to make high-strength, low-volume, large parts. Fillers and reinforcing materials can enhance shrinkage control, chemical and shock resistance, electrical and thermal insulation, and/or reduce cost.

优点

反应成型分析具有以下优点：
- 热固性的交链聚合物结构通常可提供改进的机械属性以及更强的耐热性和耐环境性。
- 热固性材料的黏度通常很低，与热塑性成型相比，用相对较低的压力和锁模力即可使大型的复杂零件成型。
- 热固性材料可用于复合工艺。例如，RTM 和 SRIM 工艺使用长纤维制成的预塑，提供了制作强度高、体积小的大型零件的方法。填充物和加固材料可增强收缩控制、耐化学性和抗冲击性、电绝缘和热绝缘，并且/或者可降低成本。

第七节 词汇表
Section 7　Words

Tab. 13-1　　　　　　　　　　Glossary of Terms　术语表

Glossary　术语	Definition　定义
Mould for plastic 塑料成型模具(塑料模)	Molding plastic parts with molds in the plastic molding process 在塑料成型工艺中,成型塑件用的模具
Mould for thermoplastics 热塑性塑料模	Mold used in the thermoplastic molding 热塑性塑料成型时使用的模具
Mould for thermoset 热固性塑料模	Thermosetting plastic molding mold 热固性塑料成型时使用的模具
Compression mould 压缩模	With heat and pressure, the mold for the plastic melt and solidify forming, directly into the cavity 借助加压和加热,使直接放入型腔的塑料熔融并固化成型所用的模具
Transfer mould 压注模 传递模	Through the plunger, in the mold, heated melting plastics thermosetting plastics in the feeding cavity, after pouring system, pressed into the heated closed cavity, solidify forming. 通过柱塞,使在加料腔内受热塑化熔融的热固性塑料,经过浇注系统,压入被加热的闭合型腔,固化成型所用的模具。
Injection mould 注射模	By the injection machine screw or a piston, so that the plasticization molten plastic in material barrel, through the nozzle, casting system, into the cavity, forming. 由注射机的螺杆或活塞,使料桶内塑化熔融的塑料,经喷嘴、浇注系统,注入型腔,固化成型所使用的模具。
Injection mould for thermoplastics 热塑性塑料注射模	The mold used for molding of thermoplastics 成型热塑性塑件所用的模具
Injection mould for thermoset 热固性塑料注射模	The mold used for molding thermoset plastic parts 成型热固性塑件所使用的模具
Flash mould 溢式压缩模	The heating chamber that cavity. Clamping, allow excessive plastic overflow in compression molding. 加热腔即型腔。合模时,允许过量的塑料溢出的压缩模。
Semi-positive mould 半溢式压缩	The heating chamber is the expansion of part of the cavity. Clamping, allows a small amount of plastic overflow in the compression molding. 加热腔是型腔上的扩大部分。合模时允许少量的塑料溢出的压缩模。
Positive mould 不溢式压缩模	The feeding chamber is a continuation of part of the up cavity. The working pressure is all applied to the plastic, almost no compression molding plastic overflow. 加料腔是型腔向上的延续部分。工作压力全部施加在塑料上,几乎无塑料溢出的压缩模。

（续表）

Glossary 术语	Definition 定义
Portable compression mould 移动式压缩模	The forming of secondary operations such as mold, unloading, loading, clamping and moved to the compressor working table of compression mold. 将成形中的辅助作业如开模、卸件、装料、合模等移到压机工作台面外进行的压缩模。
Fixed compression mould 固定式压缩模	Fixed on the compressor working table, all forming operations are in the machine tools on the compression mold. 固定在压机工作台面上，全部成形作业均在机床上进行的压缩模。
Fixed transfer mould 固定式压注模	Fixed on the compressor working table, all forming operations are in machine tools on the transfer mould. 固定在压机工作台面上，全部成形作业均在机床上进行的压注模。
Runnerless mould 无流道模	In successive forming operations, using proper temperature control, the flow channel in a melt flow state of the injection mold, including the use of extended nozzle injection mold. 在连续成形作业中，采用适当的温度控制，使流道内保持熔融流动状态的注射模，包括采用延伸喷嘴的注射模。
Hot runner mould 热流道模	Continuous forming operation, by means of heating, thermoplastic within the flow channel always maintains the melt flow state of the injection mold. 连续成形作业中，借助加热，使流道内的热塑性塑料始终保持熔融流动状态的注射模。
Insulated runner mould 绝缘热流道模	Continuous forming operations, with plastic and channel wall contact solid layer playing the role of thermal insulation, the thermoplastic in the center part of flow channel on injection mold has always maintained a molten state. 连续成形作业中，利用塑料与流道壁接触的固体层所起的绝热作用，使流道中心部位的热塑性塑料始终保持熔融状态的注射模。
Warm runner mould 温流道模	Continuous forming operation, the proper temperature control, so that the thermosetting plastic within the flow path always maintain the melt flow state on the injection mold. 连续成形作业中，采用适当的温度控制，使流道内的热固性塑料始终保持熔融流动状态的注射模。
Feed system 浇注系统	The feeding passage from the injection machine nozzle or feeding chamber of the transfer mould to the cavity, including the sprue, runner, gate and the cold slug well. 由注射机喷嘴或压注模加料腔到型腔之间的进料通道，其中包括主流道、分流道、浇口和冷料穴。

（续表）

Glossary 术语	Definition 定义
Sprue 主流道	Injection mold, the feed channel used to connect the injection machine nozzle and the cavity (single cavity mold) or runner; transfer mould, the feed channel used to connect the feeding chamber and cavity (single cavity mold) or runner. 注射模中，使注射机喷嘴与型腔（单型腔模）或与分流道连接的这一段进料通道。压注模中，使加料腔与型腔（单型腔模）或与分流道连接的这一段进料通道。
Runner 分流道	The feeding channel connecting the sprue and gate 连接主流道和浇口的进料通道
Gate 浇口	A gate is the feeding channel which connects the runner and the cavity. 连接分流道和型腔的进料通道。
Direct gate, sprue gate 直接浇口	A direct gate is commonly used for single-cavity molds, where the sprue feeds material directly and rapidly into the cavity with minimum pressure drop. 直接浇口通常用于单型腔模具，其中主流道以最小压力降快速将材料直接注入型腔。
Ring gate 环形浇口	With a ring gate, the material flows freely around the core before it moves down as a uniform tube-like extrusion to fill the mold. 使用环形浇口时，材料会先绕型芯自由流动，然后再像均匀管状成形物一样向下流动以填充模具。
Disk gate, diaphragm gate 盘式浇口	A gate is a circumferentially extending feed gate along the plastic parts. 沿塑件内圆周扩展进料的浇口。
Spoke gate, spider gate 轮辐浇口	Runner like spokes distributed in the same plane, it is a circumferentially extending feed gate is along the plastic part. 分流道像轮辐状分布在同一平面内，沿塑件的部分圆周扩展进料的浇口。
Pin-point gate 点浇口	The gate which cross section shape is as pin point. 截面形状小如针点的浇口。
Edge gate 侧浇口	An edge gate is located on the parting line of the mold. The gate cross section is rectangular and can be tapered in width and/or thickness between the part and runner. 设置在模具的分型处，浇口横截面为矩形，可以实现零件和流道之间宽度和/或厚度的锥形化。
Submarine gate, tunnel gate 潜伏式浇口	A submarine gate is similar to an edge gate but a portion of the gate overlaps the part. 潜伏式浇口与侧缘浇口相似，只是浇口的一部分重叠在零件上。

（续表）

Glossary 术语	Definition 定义
Tab gate 护耳浇口	A tab gate is typically employed for parts that require low shear stresses, such as optical parts. The high shear stress generated around the gate is confined to the auxiliary tab, which is trimmed off after molding. 耳浇口通常用于要求低剪切应力的零件，例如光学零件。浇口周围产生的高剪切应力被限制在辅助护耳内，成型后将修剪掉该护耳。
Fan gate 扇形浇口	A fan gate is a wide edge gate with variable thickness, which permits rapid filling of large parts or fragile mold sections through a large entry area. Fan gates are used to create a uniform flow front into wide parts where warping and dimensional stability are main concerns. 扇形浇口是厚度可变的宽边浇口，允许通过宽大的入口来快速填充大零件或模具易碎部位。扇形浇口用于为宽零件创建均匀的流动前沿，其中翘曲和尺寸稳定是需要考虑的两个主要方面。
Cold-slug well 冷料穴	In injection mould, cold-slug well is directly opposite the sprue hole or groove and is used to store the cold material. 注射模中，直接对着主流道的孔或槽，用以储存冷料。
Sprue bush, sprue bushing 浇口套	Sprue is a lining part with sprue channel which directly contact with injection machine nozzle or loading chamber of transfer mold. 直接与注射机喷嘴或压注模加料腔接触，带有主流道通道的衬套零件。
Gating insert 浇口镶件	In order to improve the service life of the gate, the gate could be replaced with metal insert. 为提高浇口的使用寿命，对浇口采用可更换的耐磨金属镶块。
Spreader 分流锥	Spreader is located in the sprue, used to make plastics shunt and gentle changes the flow direction, usually with a conical cylindrical part. 设在主流道内，用以使塑料分流并平缓改变流向，一般带有圆锥头的圆柱形零件。
Runner plate 流道板	Runner plate is a plate set specially for the creation of the shunt. 为开设分流道专门设置的板件。
Manifold block Hot-runner manifold 热流道板	In hot-runner mould, manifold is a board or columnar part, where the heating element is placed for shunt, and used to make the flow passage of the thermoplastic always maintain the melt flow state. 在热流道模中，为开设分流道放置加热元件，用以使流道的热塑性塑料始终保持熔融流动状态的板状或柱状零件。
Warm runner plate 温流道板	In warm runner mould, the plates where the shunt is set are referred to as warm runner plate. 在温流道模中，开设分流道的板，均称为温流道板。
Secondary nozzle 二级喷嘴	The nozzle is provided for direct of indirect feeding channel from the hot runner plate to the cavity. 为热流道板向型腔直接或间接提供进料通道的喷嘴。

(续表)

Glossary 术语	Definition 定义
Torpedo, torpedo body assembly 鱼雷型组合体	Torpedo is torpedo shape combination that set in the sprue bushing or secondary nozzle of hot runner mold, shunting and heating effects, including the torpedo head, torpedo body and cartridge heater 设置在热流道模浇口套或二级喷嘴内,起分流和加热作用的鱼雷形状的组合体,包括鱼雷头、鱼雷体和管式加热器。
Cartridge heater 管式加热器	The tubular heating elements are set in the hot runner plate or torpedo body. 设置在热流道板或鱼雷体内的管形加热元件。
Heat pipe 热管	The high efficient heat conduction component, to reduce the difference in temperature between the hot runner and gate. Also used for mold cooling system. 缩小热流道和浇口之间温差的高效导热元件。也可用于模具的冷却系统。
Valve gate 阀式浇口	The gate form, set in the secondary nozzle of the hot runner, to control the molten plastics discharge or stop flow by valve. 设置在热流道二级喷嘴内,利用阀门控制熔融塑料放流或止流的浇口形式。
Loading chamber 加料腔	Refer to the extension of (die) cavity open end, used for additional loading space, in compression mould. 在压缩模中,指(凹模)型腔开口端的延续部分,用来附加装料的空间。 The chamber body parts, used to store the plastics and heat it before it entering the (mould) cavity, in transfer mould. 在压注模中,指塑料在进入(模具)型腔前,盛放并使之加热的腔体零件。
Force plunger, pot plunger 柱塞	Cylindrical part, used to transfer the machine pressure so that the plastic in loading chamber can be injected into the feeding system and cavity in the transfer mould. 压注模中,传递机床压力,使加料腔内的塑料注入浇注系统和型腔的圆柱形零件。
Flash groove, spew groove 溢流槽	The slots opened in the mould to exclude the excess plastics in compression mould. 在压缩模中,为排除过剩的塑料而在模具上开设的槽。 Overflow trench opened in the mould to avoid to arise the weld of the plastic parts, in the injection. 在注射模中,为避免在塑件上可能产生熔接痕而在模具上开设排溢用的沟槽。
Vent 排气槽	The gas flow slots or holes, opened in the mould to discharge the cavity gas. 为使型腔内的气体排出模具外,在模具上开设的气流通槽或孔。
Parting 分型	

第十三章 塑料成型与模具制造

(续表)

Glossary 术语	Definition 定义
Parting line 分型面	The separable contact surface, used to remove the condensate material from plastic parts and (or) feeding system in the mould. 模具上用以取出塑件和(或)浇注系统凝料的可分离的接触表面。
Horizontal parting line 水平分型面(线)	The mould parting surface parallel to the surface of the worktable in the press or injection. 与压机或注射机工作台面平行的模具的分型面。
Vertical parting line 垂直分型面(线)	The mould parting surface perpendicular to the surface of worktable in the press or injection. 与压力机或注射机工作台面垂直的模具的分型面。
Mold parts-Molding parts 模具零件-成型零件	
Stationary mould fixed half 定模	The half of the mold, installed on the fixed worktable in the injection. 安装在注射机固定工作台面上的那一半模。
Movable mould moving half 动模	The half of the mold, installed on the mobile worktable in the injection for opening and closing with the injection. 安装在注射机移动工作台面上的那一半模具,可随注射机作开闭运动。
Upper mould, upper half 上模	In the compression mould and transfer mould, refer to the half of the mould installed upper worktable in the press. 在压缩模和压注模中,安装在压机上工作台面上的那一半模具。
Lower mould, lower half 下模	In the compression mould and transfer mould, refer to the half of the mould installed lower worktable in the press. 在压缩模和压注模中,安装在压机下工作台面上的那一半模具。
Cavity 型腔	Used to fill the plastic into the form space (i.e., the mold cavity), when molding. 合模时,用来填充塑料,成形塑件的空间(即模具型腔)。 Sometimes also refers to the die cavity (i.e., concave cavity). 有时也指凹模中成形塑件的内腔(即凹模型腔)。
Impression, cavity block, cavity plate 凹模	The concave parts (including cavity and entities), to form the external the surface of the plastic parts. 成形塑件外表面的凹状零件(包括零件的内腔和实体两部分)。
Mould insert 模具镶件	The local forming parts can be manufactured separately form the main body and embedded it, when forming parts (die, punch or core) having vulnerable or local position to be manufactured overall. 当成形零件(凹模、凸模或型芯)有易损或难以整体加工的部位时,与主体件分离制造并嵌在主体件上的局部成形零件。
Moveable insert, loose detail 活动镶件	The inserts can be removed separately to form the plastic parts together with drawing out of the plastic parts, according to the process and structural requirements. 根据工艺和结构要求,须随塑件一起出模,方能从塑件中分离取出的镶件。

(续表)

Glossary 术语	Definition 定义
Splits 拼块	A number of parts can be manufactured separately and used to put together to the cavity or core, based on the design and process requirements. 按设计和工艺要求，用以拼合成凹模或型芯的若干分离制造的零件。
Cavity splits 凹模拼块	A number of parts can be manufactured separately and used to put together to the cavity. 用于拼合成凹模的若干分离制造的零件。
Core splits 型芯拼块	A number of parts can be manufactured separately and used to put together to the core. 用于拼合成型芯的若干分离制造的零件。
Core 型芯	The convex part, to form the inner surface of the plastic part. 成形塑件内表面的凸状零件。
Side core, slide core 侧型芯	
Thread core, threaded core 螺纹型芯	The part, directly to form the internal thread. 直接成形塑件内螺纹的零件。
Thread ring, threaded cavity 螺纹型环	The part, directly to form thread of the plastic parts. 直接成形塑件螺纹的零件。
Punch, force 凸模	The part, directly contacting the plastic part, with segment assembly with cavity, to form the inner surface or upper and lower end surface of the part, under the pressure of the compression mould. 压缩模中，承受压机压力，与凹模有配合段，直接接触塑料，成形塑件内表面或上、下端面的零件。
Insert 嵌件	The metal or other materials, buried into the plastic parts in forming process. 成形过程中，埋入或随压入塑件中的金属或其他材料的零件。
Support and fixed parts 支撑与固定零件	
Fixed clamp plate, top clamping plate, top plate 定模具座板	The plate, to fix the stationary mold half on the stationary worktable of the injection. 使定模固定在注射机的固定工作台面上的板件。
Moving clamp plate, bottom clamping plate, bottom plate 动模座板	The plate, to fix the movable mould half on the moving worktable of the injection. 使动模固定在注射机的移动工作台面上的板件。
Upper clamping plate 上模座板	The pate, to fix the upper die to upper worktable in the press. 使上模固定在压机上工作台面上的板件。
Lower clamping plate 下模座板	The plate, to fix the lower die on lower worktable in the press. 使下模固定在压机下工作台面上的板件。

（续表）

Glossary 术语	Definition 定义
Cavity-retainer plate 凹模固定板	Plate parts, to fix the cavity. 用于固定凹模的板状零件。
Core-retainer plate 型芯固定板	Plate parts, to fix the core. 用于固定型芯的板状零件。
Punch-retainer plate 凸模固定板	Plate parts, to fix the punch. 用于固定凸模的板状零件。
Chase, bolster, frame 模套	The collection of box-shaped structural components to fix the concave mold or core. 固定凹模或型芯的框套形结构零件的统称。
Backing plate, support plate 支撑板	The plate, to prevent the molding parts (die, punch, core and inserts) and guide part to move on axial direction, and withstand the molding pressure. 防止成形零件（凹模、凸模、型芯或镶件）和导向零件轴向移动并承受成形压力的板件。
Spacer parallel 垫块	Massive parts, adjust the height of the closed mold, to form the space required by ejector mechanism. 调节模具闭合高度，形成推出机构所需的推出空间的块状零件。
Ejector housing, mould base leg 支架	The pad, to make the movable half of mould to be fixed in the press or injection. 使动模能固定在压机或注射机上的垫块。
Support pillar 支撑柱	Cylindrical parts, set between the support plate and clamping plate of the moveable mould half, playing a supporting role, to enhance the rigidity of movable mould plate. 为增强动模的刚度而设置在动模支承板和动模座板之间，起支撑作用的圆柱形零件。
Mould plate 模板	A collection of plates composited of the mould. 组成模具的板类零件的统称。
Core pulling parts 抽芯零件	
Angle pin, finger cam 斜销	Cylindrical part in the mold, assembled tilting at the parting surface to make the slider relative motion, with the mold open and close. 倾斜于分型面装配，随着模具的开闭，使滑块在模内产生相对运动的圆柱形零件。
Silder 滑块	沿导向件上滑动，带动侧型芯完成抽芯和复位动作的零件。 Sliding along the guide, the side core pulling and the reset action completed.
Side core-slide 侧型芯滑块	The side core and slider, made by the whole material. 由整体材料制成的侧型芯和滑块。

(续表)

Glossary 术语	Definition 定义
Slide guide strip 滑块导板	Guide plate, assembling with the sliding surface of the slider. 与滑块的导滑面配合,起导向作用的板件。
Heel block, wedge block, locking heel 楔紧块	The parts, with a wedge angle for wedging tight solider, when clamping. 带有楔角,用于合模时楔紧滑块的零件。
Finger guide plate 斜槽导板	Plate part, with inclined guide slot, used for core pulling and reciprocating motion. 具有斜导槽,用以使滑块随槽作抽芯和往复运动的板状零件。
Clog-leg cam 弯销	Rectangular or square cross-section of the curved rod parts, with the mold opening and closing, to make the core pulling and resetting action. 矩形或方形截面的弯杆零件,随着模具的开闭,使滑动抽芯,复位动作。
Angled-lift splits 斜滑块	Splits, used to forming, pushing and core pulling, assembling with the inclined plane to slide. 利用与斜面配合而产生滑动,兼有成形、推出和抽芯作用的拼块。
Guide part 导向零件	
Guide pillar, guide pin, leader pin 导柱	Cylindrical parts, assembling with the guide sleeve (or hole) in the other half mould, used to determine the relative position of movable and stationary mold. In order to ensure the accuracy of mold movement. 与安装在另一半模上的导套(或孔)相配合,用以确定动、定模的相对位置。保证模具运动导向精度的圆柱形零件。
Guide pillar straight, straight leader pin 带头导柱	Guide post, with axial positioning step, and the fixed segment and guiding segment with the same nominal size, and with different tolerances. 带有轴向定位台阶,固定段与导向段具有同一公称尺寸,不同公差带的导柱。
Guide pillar, shouldered, shoulder leader pin 带肩导柱	Guide post, the nominal size of fixed segment is greater than the guiding segment, with axial position step. 带有轴向定位台阶,固定段公称尺寸大于导向段的导柱。
Ejector guide pillar, ejector guide pin 推板导柱	Cylindrical part, sliding with the guide bush of ejector plate, used to guide the ejector mechanism. 与推板导套滑配合,用于推出机构导向的圆柱形零件。
Guide bush, guide bushing 导套	Round dwelling pat, assembling with the guide pin installed in the other half of the mold used to determine the relative position of movable and stationary molds and to ensure the moving precise of the mold. 与安装在另一半模上的导柱相配合,用以确定动、定模的相对位置,保证模具运动导向精度的圆套形零件。
Guide bush, straight, straight bushing 直导套	The guide sleeve, without axial positioning step. 不带轴向定位台阶的导套。

第十三章 塑料成型与模具制造

（续表）

Glossary 术语	Definition 定义
Guide bush, shoulder bushing 带头导套	The guide sleeve, with axial positioning step. 带有轴向定位台阶的导套。
Ejector guide bush, ejector bushing 推板导套	Cylindrical part, sliding with the push plate guide pin to guide the ejector mechanism. 与推板导柱滑配合，用于推出机构导向的圆形零件。
Positioning and limit parts 定位和限位零件	
Locating ring 定位圈	The locating part, alimenting the nozzle of ejector mechanism and sprue bush of mould, to decide the position of mould on the ejector mechanism. 使注射机喷嘴与模具浇口套对中，决定模具在注射机上安装位置的定位零件。
Mould bases locating elements 锥形定位件	The components, when mould clamping, corresponding with the cone to locate the moveable and stationary halves of mould to precise positions. 合模时，利用相应配合的锥面，使动、定模精确定位的组件。
Ejector plate return pin, push-back pin 复位杆	The rod, with the closing action of the mold, to reset the ejector mechanism. 借助模具的闭合动作，使推出机构复位的杆件。
Length bolt, puller bolt 定距拉杆	A bar, used to limit a mould plate, with open and stop action only in the limit distance, when mould is opening. 在开模分型时，用来限制某一模板，仅在限定的距离内作拉开和停止动作的板件。
Puller plate, limit plate 定距拉板	A plate, used to limit a mould plate, with open and stop action only in the limit distance, when mould is opening. 在开模分型时，用来限制某一模板，仅在限定的距离内作拉开和停止动作的板件。
Ejector parts 推出零件	
Ejector pin 推杆	A bar is used to push out plastic part or condensation material in feeding system. 用于推出塑件或浇注系统凝料的杆件。
Ejector pad 推块	Part forming effect in the cavity, massive parts, pushing out the plastic part from cavity when mould is opening. 在型腔内起部分成形作用，并在开模时把塑件从型腔内推出的块状零件。
Stripper plate 推件板	Plate-shaped part, directly pushing out the plastic part. 直接推出塑件的板状零件。
Stripper ring, stripper disk 推出环（盘）	Ring or disk parts, pushing out the plastic part locally or entirely. 起局部或整体推出塑件作用的环形或盘形零件。

(续表)

Glossary 术语	Definition 定义
Ejector retainer plate 推杆固定板	Plate used for retaining the ejector and reset parts and ejector guide bush. 用以固定推出和复位零件的以及推板导套的板件。
Ejector plate, ejection plate 推板	Supporting roll and reset parts, direct transfer machine roll force. 支撑推出和复位零件,直接传递机床推出力的板件。
Ejector tie rod 连接推杆	A bar is connected between the ejector plate and ejector-retainer plate, in order to transfer force. 连接推板与推杆固定板,传递推力的杆件。
Sprue puller 拉料杆	In order to pull out the condensation material in sprue bush, a bar is set with grooves or other shapes in head, opposite the sprue. 为了拉出浇口套内的浇注凝料,在主流道的正对面,设置头部带有凹槽或其他形状的杆件。
Runner plate, runner lock pin 分流道拉料杆	A bar is introduced to ensure to hold the condensation material in runner temporarily; whose one end buried in the runner is formed by inverted cone shape or other shape. 将埋入分流道的一端制成倒锥形或其他形状,用以保证开模时暂时拉住分流道凝料的杆件。
Runner stripper plate 推流道板	Panel is introduced to push out the condensation material in feeding system, along with the mould opening. 随着开模运动,推出浇注系统凝料的板件。
Cooling and heating parts 冷却和加热零件	
Cooling channel, cooling line 冷却通道	Within the mold through the cycle of cooling water or channel of other medium, in order to control the requirements of the mold temperature. 模具内通过冷却循环水或其他介质的通道,用以控制所要求的模具温度。
Baffle 隔板	In order to change steam or cooling water flow, the metal strips or plates are arranged in the mold cooling channel. 为改变蒸汽或冷却水的流向而在模具的冷却通道内设置的金属条或板。
Heating plate 加热板	The structure plate is set up by hot water, steam or electric, in order to ensure mold plastic parts forming temperature requirements. 为保证模具内塑件成形的温度要求而设置的热水、蒸汽或电等加热结构的板件。
Thermal insulation board 隔热板	A plate in which the heat transfer is prevent. 防止热量传递的板件。
Mould base 模架	The combination consists of mold plate, guide pillar and guide bush etc., but the cavity is unprocessed. 由模板、导柱和导套等零件组成,但型腔未加工的组合体。

第十三章 塑料成型与模具制造

（续表）

Glossary 术语	Definition 定义
Standard mould base 标准模架	The mold base is combined by a set of certain interchangeability parts, which are standardized, serialized on the structure, form and dimensions. 由结构、形式和尺寸都标准化、系列化并具有一定互换性的零件成套组合而成的模架。
Main design factors for plastic molding 塑料成型的主要设计要素	
Shot capacity 注射能力	In a molding cycle, the maximum injection capacity or weight given by the injection molding machine. 在一个成形周期中，注射机对给定塑料的最大注射容量或重量。
Shrinkage 收缩率	At room temperature, the ratio of difference of the two linear dimensions between the mold cavity and corresponding plastic parts and the parts or molds origin linear dimensions. 在室温下，模具型腔与对应塑件二者的线性尺寸之差和对塑件或模具线性尺寸之比。
Injection pressure 注射压力	The pressure is exerted by the injection molding machine, when the molten plastic is injected to mold cavity. 注射机使熔融塑料注入模具型腔时所需施加的压力。
Clamping force, locking force 锁模力	Forming process, the force is exerted on the mold, in order to ensure the fixed mold and moveable mold to be closed tightly to each other. 成形过程中，为保证动、定模相互紧密闭合而需施加在模具上的力。
Molding pressure 成型压力	Press to exert per unit area pressure on the projected area of the plastic parts. 压机施加在塑件单位投影面积上的压力。
Internal mould pressure, cavity pressure 模内压力	The pressure on the cavity surface given by molten plastic in the injection pressure. 在注射压力下的熔融塑料对型腔表面的压力。
Mould opening force 开模力	After forming, the required force to separate the mould from parting line. 成形后，使模具从分型面上分开所需的力。
Ejection force 脱模力	The required force that plastic part is emerged from the mold. 使塑件从模内脱出所需的力。
Core-pulling force 抽芯力	The force required, when the molding plastic part drawn and pulled out the side of the core. 从模内的成形塑件中，抽拔出侧型芯所需的力。
Core-pulling distance 抽芯距	The moving distance of side core or slider required, when side core drawn from the molding position without prejudice to the plastic parts, to removing location. 将侧型芯从成形位置抽至不妨碍塑件取出位置时，侧型芯或滑块所需移动的距离。

(续表)

Glossary 术语	Definition 定义
Mould shut height 闭合高度	The total height, mold in the closed state. 模具处于闭合状态下的总高度。
Maximum daylight, open daylight 最大开距	The maximum distance separated between stationary and moveable or upper and lower worktable in injection machine or press. 注射机或压机的动、定工作台或上、下工作台之间可分开的最大距离。
Projected area 投影面积	Compressive mold, in the projection plane perpendicular to the direction of pressure, the total area of the feeding chamber projection. 压缩模中,在与加压方向垂直的投影面上,加料腔投影的总面积。 Injection mold, in the projection plane perpendicular to the direction of the clamping force, the total area of injection plastic projection. 注射模中,在与锁模力方向垂直的投影面上,注射塑料投影的总面积。
Draft 脱模斜度	The angle included in the part design to enable the part to be easily ejected from the mold. 在零件设计中包括的角度,可使零件轻松地从模具中顶出。
Stripper distance 脱模距	After parting, the distance is required to remove the plastic pieces and condensate material in sprue and runner. 分模后,取出塑件和主、分流道凝料所需的距离。

第十四章 机械工程材料
Chapter 14　Material in Mechanical Engineering

第一节　工程材料的性能
Section 1　Property of Engineering Material

- ★ 性能　ability/performance
- ★ 材料　material
- 分类　assorting
- 弹性变形　elastic deformation
- 弹性模量　elasticity modulus/elastic modulus
- 弹性极限　elastic limit
- 弹性　elasticity
- ★ 塑性　plastic behavior
- ★ 强度　intensity/strength
- 抗拉强度　strength for extension/strength of extension
- ★ 韧性　toughness/tenacity
- ★ 硬度　rigidity/hardness
- 耐磨性　abrasive resistance
- 真实应力　actual stress
- 交变应力　alternate stress/alternating stress
- 截面积　cross section
- 布氏硬度　ball-pressure hardness/ball-thrust hardness/brinell hardness
- 布氏硬度值　brinell figure/brinell [hardness] number
- 布氏硬度试验　brinell test of hardness/brinell hardness test
- 断裂/断口　break/breakdown
- 脆性　brittleness
- 脆性破坏　brittle break/brittle failure
- 脆[性断]裂　brittle fracture
- 脆性转变温度　brittle transition temperature

There are more than 50,000 materials available to the engineer. Engineering design, then, involves many considerations. All materials have their own properties or characteristics. These properties include: physical properties, mechanical properties, chemical properties, thermal properties, electrical and magnetic properties, optical properties, and acoustical properties. The engineering materials provide good service to human by their quality properties. The choice of a material must meet certain criteria on bulk and surface properties (e.g. strength and corrosion resistance). But it must also be easy to fabricate; it must appeal to potential consumers; and it must compete economically with other alternative materials.

有超过 50 000 种的材料可供工程师进行选择。因而工程师在工程设计中需要从多方面进行比较和考虑。任何材料都有其性能或特性，这些性能主要包括：物理性能、机械性能、化学性能、热学性能、电磁性能、光学性能和声学性能。工程材料是通过它们所具有的优良性能服务于人类的。选择材料必须符合特定的体积和表面特性的标准（如强度和耐腐蚀）。同时材料也必须适于加工，与其他材料相比在成本上也要有足够的竞争力。

第二节　材料结构
Section 2　Material Structure

- ★ 合金　alloy
- 成分/组元　component
- 晶格　crystalline host lattice
- ★ 晶体结构　crystal structure
- ★ 体心立方结构　body-centered cubic crystal structure
- ★ 面心立方结构　face-centered cubic crystal structure
- ★ 密排六方结构　close-packed hexagonal structure
- 晶体位向,晶体取向　crystal orientation
- 溶质原子　solute atom
- 固溶体　solid solution
- 金属化合物　metallic compound
- ★ 晶粒　grain/crystal grain
- 晶面　crystal plane

晶向	crystal orientation	原子排列	atomic arrangement
晶粒度	grain size	★ 位错	dislocation
粗晶粒	coarse grain	刃型位错	edge dislocation
细晶粒	fine grain	原子位错,原子位移	atomic dislocation
晶界	grain boundary	晶体缺陷	crystal defect
各向异性	anisotropies, anisotropy	晶胞	cell

It should be clear that all matter is made of atoms. These atoms form thousands of different substances ranging from the air we breathe to the metal used to support tall buildings. All metals, many ceramic materials, and certain polymers are in the form of crystalline structures under normal solidification conditions. For a crystalline solid we have tacitly assumed that perfect order exists throughout the material on an atomic scale. However, such an idealized solid does not exist; all contain large numbers of various defects or imperfections. As a matter of fact, many of the properties of materials are profoundly sensitive to deviations from crystalline perfection; the influence is not always adverse, and often specific characteristics are deliberately fashioned by the introduction of controlled amounts or numbers of particular defects, as detailed in succeeding chapters.

所有的物质都是由原子组成的。这些原子形成了成千上万的不同物质,从我们呼吸的空气到用于支撑高楼的金属。所有的金属材料和大多数的陶瓷材料以及某几种高分子材料,在正常凝固状态下都具有晶体结构。对于一个晶体,我们都假定其原子尺度上的结构是完美的。但是,这样理想化的晶体是不存在的,晶体中都含有大量缺陷或瑕疵。事实上,许多材料的晶体结构并不完美而是有缺陷的,而这些缺陷对应用并非总有不利影响,有时会人为引入某些特定形式的缺陷。

第三节 材料的凝固
Section 3　Material Solidification

凝固	solidification	★ 过冷	supercooling, undercooling
固化	solidify	枝晶长大	dendrite growth
★ 结晶	crystallization	枝(状)晶	dendrite
★ 成核(形核)	nucleation	柱状晶	columnar crystal
过冷度	degree of supercooling	原子团	aggregate/aggregate of atoms
晶核	crystal nucleus	同素异构	allotrope
★ 晶粒长大	crystal grain growth	★ 同素异构转变	allotropic transition

1 In production engineering, metallurgy is concerned with the production of metallic components for use in consumer or engineering products. This involves the production of alloys, the shaping, the heat treatment and the surface treatment of the product. The task of the metallurgist is to achieve balance between material properties such as cost, weight, strength, toughness, hardness, corrosion, fatigue resistance, and performance in temperature extremes.

在工业生产中,冶金为消费者或工程产品提供金属组件。这涉及合金、成型、热处理和表面处理等加工。冶金学研究人员的任务是实现材料性能之间的平衡,如成本、质量、强度、韧性、硬度、耐腐蚀、抗疲劳以及极端温度下的性能。

2 Usually, solidification is a basis segment for preparation material. The crystallization process consists of two major events, nucleation and crystal growth.

通常,凝固是材料制备的基本手段。结晶过程包括两个主要过程,即成核和晶体生长。

第四节 二元相图及其应用
Section 4　Binary Phase Diagram and Its Application

★ 相图	phase diagram	二元相图	binary phase diagram/binary-component phase diagram
二元合金	binary alloy		

第十四章 机械工程材料

杠杆定律(相图)　centre of gravity method
★ 冷却曲线　cooling curve
★ 铁素体　ferrite
★ 珠光体　pearlite
★ 奥氏体　austenite
莱氏体　ledeburite
单相组织　single-phase structure
两相组织　two-phase structure
★ 共晶反应　eutectic reaction
★ 共析反应　eutectoid reaction
★ 渗碳体　cementite
α-铁素体　alpha-ferrite
网状析出　cellular precipitation
渗碳体网状组织、网状渗碳体　cementite network
显微组织变化、显微结构变化　change in microstructure
★ 相变　change of phase
相变点、转变点　change point

A phase diagram in physical chemistry, engineering, mineralogy, and materials science is a type of chart used to show conditions at which thermodynamically distinct phases can occur at equilibrium. The understanding of phase diagrams for alloy systems is extremely important because there is a strong correlation between microstructure and mechanical properties, and the development of microstructure of an alloy is related to the characteristics of its phase diagram.

相图是一个用来在物理化学、工程学、矿物学和材料科学上表示热条件不同的情况下平衡状态的图表。了解相图对于了解合金系非常重要，因为合金的显微结构和机械性能之间有非常紧密的联系，并且相图的特征与合金的显微结构也有非常密切的关系。

第五节　材料的变形
Section 5　Material Deformation

解理　cleavage
解理断裂　cleavage crack/cleavage fracture
韧性断裂　ductile fracture
塑性　plastic behaviour/plasticity
塑性变形　plastic deformation
变形　deformation
形变带　deformation band
滑移方向　direction of slip
滑移线　sliding line/slip line
热加工　heat deformation
冷加工　cold deformation
回复　come-back
★ 再结晶　recrystallization
绝对温度　absolute temperature
纤维状组织　fibrous structure
变形织构　deformation texture
★ 加工硬化　cold hardening/cold quenching
临界分切应力　critical resolved shear stress
临界应力　critical stress
超塑性　superplasticity

Materials and metallurgical engineers, on the other hand, are concerned with producing and fabricating materials to meet service requirements as predicted by these stress analyses. This necessarily involves an understanding of the relationships between the microstructure (i. e., internal features) of materials and their mechanical properties. Plastic working is a process in which the work-piece is shaped by compressive forces through various dies and tools. Its principle is plastic deformation.

材料和冶金工程师关心的是生产和加工达到所需应力要求的材料。因此一定要了解材料的微观组织(例如内部特性)与力学性能之间的关系。压力加工是一种通过各种模具和工具利用压力使工件成形的工艺方法。它的原理就是塑性变形。

第六节　钢的热处理
Section 6　Heat Treatment of Steel

加热　heating
转变　transformation
冷却　cooling
冷却转变　athermal
★ 热处理　heat treatment
相　phase
★ 相变点　transformation temperature/critical point
★ 马氏体　martensite

贝氏体　bainite
贝氏体转变　bainite transformation
针状马氏体　acicular martensite
奥氏体　austenite
残余奥氏体　retained austenite
过冷奥氏体　undercooled austenite
奥氏体分解　austenite
奥氏体转变　austenite transformation
奥氏体晶粒度　austenite grain size/austenitic grain size
奥氏体转变曲线　austenitic transformation curve
奥氏体化　austenitising/austenization/austenizing
索氏体　sorbite
托氏体　troostite
球化体（球状珠光体）　spheroidite
回火马氏体　tempered martensite
回火索氏体　tempered sorbite
回火托氏体　tempered troostite
完全奥氏体化　complete austenitizing
保温时间　holding time/soaking time
普通热处理、常规热处理　conventional heat treatment/customary heat treatment
★退火　annealing/anneal
完全退火　complete annealing/dead-soft annealing

不完全退火　annealing slack
等温退火　austenite annealing
退火处理　annealing treatment/austennealing
★正火　normalizing
★淬火　quench hardening/quenching
等温淬火　austemper/austempering
★回火　tempering/abate
时效　ageing
时效处理　ageing treatment
时效硬化　age hardening
★调质　quenching and tempering
表面热处理　surface heat treatment
硬化层深度　case depth
感应加热淬火　induction hardening
化学热处理　thermo-chemical treatment/chemical-thermal treatment
渗碳　carburizing/acieration
渗碳层　carburized case/carburized layer
渗氮（氮化）　nitriding/azotize
碳氮共渗　carbonitriding/carbo-nitriding cyaniding
残余应力　residual stress
[奥氏体]形变热处理　ausforming
时效硬化　age hardening
★连续冷却转变曲线　continuous cooling transformation(CCT) curve

Materials of all types are often heat-treated to improve their properties. Heat treatment is the operation that heats and cools a metal in its solid state to change its physical properties. The phenomena that occur during a heat treatment almost always involve atomic diffusion. Often an enhancement of diffusion rate is desired; on occasion measures are taken to reduce it.

各种类型的材料往往都可以通过热处理来提高它们的性能。热处理是在固态下加热和冷却金属从而改变其物理性能的一种工艺。热处理中几乎都存在原子扩散。一般在热处理中希望扩散速度越快越好，也有一些情况下需要减慢扩散的速度。

第七节　工业用钢
Section 7　Steel for Industry

合金元素　alloying element/alloying metal
化学成分　chemical composition
碳　carbon
碳化物　carbide/carbonide
碳化物形成元素　carbide former
合金钢　alloy steel
冷脆　cold-short
★结构钢、构件用钢　constructional steel
★碳钢、碳素钢　carbon steel
★碳素工具钢　tool steel
★调质钢　heat refined steel

★工具钢　tool steel
刃具钢　steel
冷处理　cold treatment
渗碳钢　carbon carburizing steel/carburizing steel/converted steel
模具钢　die steel
合金工具钢　alloy-tool steel
强化　strengthening
弥散硬化　dispersed phase hardening/dispersion hardening
尺寸稳定性　dimensional stability

★ 奥氏体不锈钢　austenitic stainless steel
应力腐蚀，晶间开裂　corrosion cracking
晶界腐蚀　crystal boundary corrosion
滚动轴承钢　ball bearing steel
蠕变　creep
蠕变极限　creep limit

Ferrous alloys—those of which iron is the prime constituent—are produced in larger quantities than any other metal type. They are especially important as engineering construction materials. Steels are iron-carbon alloys that may contain appreciable concentrations of other alloying elements; there are thousands of alloys that have different compositions and/or heat treatments. Alloy steel owes its properties to the presence of one or more elements other than carbon, namely nickel, chromium, manganese, molybdenum, tungsten, silicon, vanadium, and copper.

铁合金比其他任何金属材料使用得都多，尤其在工程建筑材料领域。钢是铁碳合金，也可能含有其他合金元素；有成千上万种的合金钢，它们由不同的合金元素组成或进行不同的热处理。合金钢的性质取决于其所含的除碳以外的一种或几种元素，如镍、铬、锰、钼、钨、硅、钒和铜。

第八节　铸铁
Section 8　Cast Iron

铸钢　cast steel
★ 铸铁　casting pig/cast iron
合金铸铁　alloy cast iron
石墨　black lead/graphite
石墨化　graphitizing/graphitization
★ 片状石墨　flake graphite
★ 球状石墨　spheroidal graphite
★ 团絮状石墨　temper carbon
★ 灰口铸铁　grey (pig) cast iron
麻口铸铁　mottled cast iron
添加剂　addition material
白口铸铁　galvanized iron/white cast iron
★ 球墨铸铁　nodular cast iron/spheroidal graphite cast
蠕墨铸铁　vermicular cast iron
可锻铸铁　ductile cast iron/malleable cast iron
黑心可锻铸铁　black-heart malleable cast iron

Generically, cast irons are a class of ferrous alloys with carbon contents above 2.1wt%; in practice, however, most cast irons contain between 3.0 and 4.5 wt% and, in addition, other alloying elements. A reexamination of the iron-carbide phase diagram reveals that alloys within this composition range become completely liquid at temperatures between approximately 1 150 ℃ and 1 300 ℃ (2 100 F and 2 350 F), which is considerably lower than that for steels. Thus, they are easily melted and amenable to casting. Furthermore, some cast irons are very brittle, and casting is the most convenient fabrication technique.

通常，铸铁是一种含碳量在2.1wt%以上的铁碳合金，但实际上，大多数铸铁的含碳量为3.0wt% ~ 4.5wt%，另外再加一些合金元素。铁碳合金相图显示，合金在1 150℃到1 300℃ 之间(2 100 F到2 350 F之间)完全变为液体，这远远低于钢的温度。因此，它很容易熔化铸造。此外，一些铸铁脆性较强，铸造是最合适的制造技术。

第九节　有色金属及其合金
Section 9　Non-ferrous Metal and Alloy

铝　aluminum（Al）
★ 铝合金　aluminum alloy
铜　copper/cuprum
★ 铜合金　copper alloy
完全有序　complete ordering
无序-有序转变　disorder-order transitions
青铜　bronze/bell metal
锡青铜　stannum bronze
铝青铜　aluminum bronze
铍青铜　beryllium bronze
滑动轴承　sliding bearing
巴比合金/巴氏合金　babbit/babbit alloy/babbit-metal
钛　titanium
钛合金　titanium alloy

Metal alloys, by virtue of composition, are often grouped into two classes—ferrous and non-ferrous. Although ferrous alloys are used in the majority of metallic applications in current engineering designs, non-ferrous alloys play a large and indispensable role in our technology. This section shall briefly list the major families of non-ferrous alloys and their key attributes.

金属合金按成分通常分为两类：铁碳合金和有色金属。尽管当今的工程设计所使用的金属大多数是铁碳合金，但是在我们的工程技术中，有色金属也扮演着不可缺少的角色。本节将简要介绍常用的有色金属及合金和它们的主要特性。

第十节 常用非金属材料
Section 10　Nonmetallic Material

高分子材料　macromolecule material	传统陶瓷，普通陶瓷　common pottery and porcelain
橡胶　rubber	特种陶瓷　special pottery and porcelain
合成橡胶　synthetic rubber	金属陶瓷　cermet/ceramal
合成纤维　synthetic fibre	复合材料　composite material
合成黏胶剂　synthetic mucilage glue	金属/塑料复合材料　plastimets
环氧树脂　epoxy resin	增强　gather head
酚醛树脂　phenolic resin	
★陶瓷　pottery and porcelain	

Polymers, ceramics and composite are more and more applied to all kinds of engineering.

Ceramics are compounds between metallic and nonmetallic elements; they are most frequently oxides, nitrides, and carbides. The wide range of materials that falls within this classification includes ceramics that are composed of clay minerals, cement, and glass. These materials are typically insulative to the passage of electricity and heat, and are more resistant to high temperatures and harsh environments than metals and polymers. With regard to mechanical behavior, ceramics are hard but very brittle.

Polymers include the familiar plastic and rubber materials. Many of them are organic compounds that are chemically based on carbon, hydrogen, and other nonmetallic elements; furthermore, they have very large molecular structures. These materials typically have low densities and may be extremely flexible.

A number of composite materials have been engineered that consist of more than one material type. Fiberglass is a familiar example, in which glass fibers are embedded within a polymeric material. Fiberglass acquires strength from the glass and flexibility from the polymer. Many of the recent material developments have involved composite materials.

高分子材料、陶瓷材料和复合材料越来越多地应用于各类工程中。

陶瓷是介于金属和非金属元素之间的化合物，它们是最常见的氧化物、氮化物、碳化物。在工业生产中广泛应用的材料中，属于这一分类的包括黏土材料、水泥和玻璃。这些材料通常是绝缘和隔热的，与金属和聚合物相比更耐高温和严酷环境。在机械性能上，陶瓷更硬但很脆。

聚合物包括常见的塑料和橡胶材料。其中很多是基于碳、氢和其他非金属元素的有机化合物。此外，它们有非常大的分子结构。这些材料通常具有较低的密度并且可能是很柔软的。

工程中应用的许多复合材料往往包含不止一种类型的材料。一个常见的例子是玻璃纤维，它将玻璃纤维丝嵌入到聚合物材料中。玻璃纤维从玻璃中获得足够的强度，从聚合物中获得柔韧性。最近的许多先进的材料都涉及复合材料。

第十一节 新型材料
Section 11　New Materials

形状记忆合金　shape memory alloy(SMA)	超塑性合金　superplastic alloy
形状记忆效应　shape memory effect(SME)	纳米材料　nano materials/nanometer materials
非晶态合金　amorphous alloy	

New materials refer to materials which are made with the new preparation technology or are developing; these materials have more excellent special performance than conventional material.
新型材料是指以新制备工艺制成的或正在发展中的材料,这些材料具有比传统材料更优异的特殊性能。

第十二节 工程材料的选用
Section 12　Choice of Engineering Material

★ 选材	select material	碳当量	carbon equivalent
典型	typical case	★ 强化	intensify/strengthening
实例	example/instance	★ 韧化	toughening

Properly selecting material is a consequential tache in the designing process. Many times, a material problem is one of selecting the right material from the many thousands that are available. There are several criteria on which the final decision is normally based.

First of all, the in-service conditions must be characterized, for these will dictate the properties required of the material. On only rare occasions does a material possess the maximum or ideal combination of properties. Thus, it may be necessary to trade off one characteristic for another. The classic example involves strength and ductility; normally, a material having a high strength will have only a limited ductility. In such cases a reasonable compromise between two or more properties may be necessary.

A second selection consideration is any deterioration of material properties that may occur during service operation. For example, significant reductions in mechanical strength may result from exposure to elevated temperatures or corrosive environments.

Finally, probably the overriding consideration is that of economics: what will the finished product cost? A material may be found that has the ideal set of properties but is prohibitively expensive. Here again, some compromise is inevitable.

正确地选择材料是零件设计过程中的重要环节。很多时候,选材的问题就是如何从数千种可用材料中选择合适的材料。有几种标准可以用来决定最后选择何种材料。

首先,必须满足使用条件。只有极少数情况下一种材料所有的属性都是最理想的。这样有必要主要考虑材料的一个特性并牺牲另一个。典型的例子如强度和延展性;通常,高强度的材料往往延展性有限。在这种情况下,在两个或更多属性中做一下折中是合理的。

第二个需要考虑的条件是在使用中材料失效的情况。例如,有些材料由于暴露于高温或者腐蚀性环境而造成机械强度显著下降。

最后,最重要的是经济上的考虑:即将完成的产品成本有多高?可能发现一种材料是理想的但是过于昂贵。同样不可避免地要做一些妥协。

第十五章 金属工艺学
Chapter 15　Metal Technology

第一节　金属材料基本知识
Section 1　Basic Knowledge of Metal Material

晶粒　grain	托氏体　troostite
晶界　grain boundary	球化体（球状珠光体）　spheroidite
铁素体　ferrite	回火马氏体　tempered martensite
珠光体　pearlite	回火索氏体　tempered sorbite
奥氏体　austenite	回火托氏体　tempered troostite
莱氏体　ledeburite	奥氏体化　austenitizing
单相组织　sing-phase structure	保温时间　holding time/soaking time
两相组织　two-phase structure	★退火　annealing
共晶反应　eutectic reaction	★正火　normalizing
共析反应　eutectoid reaction	★淬火　quench hardening/quenching
再结晶　recrystallization	★回火　tempering
渗碳体　cementite	时效处理　ageing treatment
热处理　heat treatment	调质　quenching and tempering
相　phase	表面热处理　surface heat treatment
相变点　transformation temperature/critical point	感应加热淬火　induction hardening
马氏体　martensite	化学热处理　thermo-chemical treatment
贝氏体　bainite	★渗碳　carburizing
残余奥氏体　retained austenite	渗碳层　carburized case
过冷奥氏体　undercooled austenite	渗氮（氮化）　nitriding
索氏体　sorbite	碳氮共渗　carbonitriding

Metals are usually malleable, ductile and shiny, that is, they reflect most of incident light. Application of metal material is very popular in industry. Materials of all types are often heat treated to improve their properties.

金属通常具有可塑性、韧性和因可以反射入射光的大部分而具有的金属光泽。金属材料在工业中的应用非常广泛。各种类型的材料往往都可以通过热处理来提高它们的性能。

第二节　铸造
Section 2　Foundry

★铸造　foundry/founding/casting	充型能力　mold-filling capacity
★铸件　casting	铸造缺陷　casting defect
造型　molding	起模斜度　pattern draft
制芯　core making	浇铸系统　running system/pouring system
浇注　pouring	直浇道　sprue
★砂型铸造　sand casting process	横浇道　runner
特种铸造　special casting process	内浇道　ingate
铸造组织　cast structure/as-cast structure	冒口　riser
铸造性能　cast ability	砂箱　flask
流动性　fluidity	芯盒　core box

砂型	sand mold	微振压实造型	vibratory squeezing molding/compaction molding
型腔	mold cavity		
分型面	mold joint/parting face	熔模铸造	lost wax casting
两箱造型	tow-part molding	消失模铸造	lost pattern casting
地坑造型	pit molding	金属型铸造	permanent mold casting
刮板造型	sweep molding	压力铸造	die casting
机器造型	machine molding	离心铸造	centrifugal pressure casting
自动化造型	automatic molding	精密铸造	precision casting

Casting is a manufacturing process in which molten metal is poured or injected and allowed to solidify in suitably shaped mold cavity. During or after cooling, the cast part is removed from the mold and then processed for delivery. Casting techniques are employed when (1) the finished shape is so large or complicated that any other method would be impractical, (2) a particular alloy is so low in ductility that forming by either hot or cold working would be difficult, and (3) in comparison to other fabrication processes, casting is the most economical. Furthermore, the final step in the refining of even ductile metals may involve a casting process.

铸造是一种制造零件的工艺，它将液态金属注入事先制备好的铸型腔内，在冷却期间或冷却之后，将铸件从铸型中取出送去进行切削加工。铸造技术在如下场合使用：(1)需要制造的零件尺寸过大或形状非常复杂，用任何其他方法加工不现实；(2)一些特殊合金延展性非常低，热加工或冷加工都是很困难的；(3)与其他制造方法相比，铸造是最经济的。此外，金属提炼的最后一步有时也会用铸造加工。

第三节　压力加工
Section 3　Press Work

★锻压	forging and stamping	压钳口	tongs hold/bar hold
金属塑性加工	plastic working of metal/metal technology of plasticity	锻造比	forging ratio
		精密锻造	precision forging/net shape forging
变形抗力	deformation stress	模锻	die forging, drop forging
可锻性	forgeability	胎模锻	loose tooling forging
各向同性	isotropy	预锻	preforging/blocking
各向异性	anisotropy	终锻	finish-forging
★加工硬化	strain hardening	★冲压	stamping/pressing
冷变形硬化	cold deformation	★板料成形	sheet forming
超塑性	superplasticity	冲压	stamping
锻造流线	forging flow line	冲裁	blanking
锻件图	forging drawing	落料	dropping
自由锻	open die forging/flat die forging	冲孔	punching
镦粗	upsetting	拉深	drawing
拔长	drawing out, swaging	拉深系数	drawing coefficient
弯曲	bending	轧辊	roller
扭转	twisting	辊锻	roll forging
切割	cutting	拉拔	drawing
压肩	necking	冷拔	cold drawing
倒棱	chamfering	挤压	extrusion

Forming operations are those in which the shape of a metal piece is changed by plastic deformation; for example, forging, rolling, extrusion, and drawing are common forming techniques. Of course, the deformation must be induced by an external force or stress, the magnitude of which must exceed the yield strength of the material. Most metallic materials are especially amenable to these procedures, being at least moderately ductile and capable of some permanent deformation without cracking or fracturing. When deformation is achieved at a temperature above that at which recrystallization occurs, the process is termed hot working; otherwise, it is cold working. For hot-working operations, large deformations are possible, which may be successively repeated because the metal remains soft and ductile.

Forging is a process in which the workpiece is shaped by compressive forces through various dies and tools. Forging is mechanically working or deforming a single piece of a normally hot metal; this may be accomplished by the application of successive blows or by continuous squeezing.

那些使金属件发生塑性变形的加工，例如锻造、滚轧、挤压和拉拔是常见的成形技术。当然，引起变形的外部力量或压力大小必须超过材料的屈服强度。大多数金属材料都适合用这些方法加工，至少中等韧性的金属材料加工后的零件可以保证不开裂。当变形在再结晶温度之上获得时，这个变形加工称为热加工，否则是冷加工。对于热加工操作，可以实现大变形，因为金属在加热后具有较高的塑性和韧性，可以重复进行加工。

锻造是一种通过各种各样的模具和工具，利用压力使工件成形的工艺方法。锻造加工是利用机械力在高温下对金属进行连续打击或挤压的过程。

第四节 焊接
Section 4 Welding

★ 焊接　welding
焊接工件　weld assembly
热影响区　heat affected zone (HAZ)
熔合区　fusion zone
熔池　molten pool/puddle
★ 焊缝　weld joint
★ 焊接接头　welding joint
★ 焊接应力　welding stress
焊后热处理　postwelding heat treatment
焊接性　weld-ability
焊接变形　welding distortion/welding deformation
焊接性试验　weldability test
★ 焊条　covered electrode
焊芯　core wire
药皮　coating
稳弧剂　arc stabilizer
熔（化）焊　fusion welding
焊接电弧　welding arc
★ 电弧焊　arc welding
★ 手工焊　manual welding
气体保护焊　gas shielded arc welding (GMAW)

氩弧焊　argon shielded arc welding
二氧化碳气体保护焊　CO_2 shielded arc welding
埋弧焊　submerged arc welding
堆焊　surfacing
电渣焊　electro-slag welding
激光焊　laser beam welding
等离子弧焊　plasma arc welding
气焊　gas welding
氧乙炔焊　oxy-acetylene welding
坡口　groove
压（力）焊　pressure welding
电阻焊　resistance welding
摩擦焊　friction welding
电阻对焊　upset welding
闪光对焊　flash welding
（电阻）点焊　resistance spot welding
爆炸焊　explosion welding
钎焊　brazing/soldering
软钎焊　soldering
硬钎焊　brazing

Welding is a kind of non-knock-down joint manner. In a sense, welding may be considered to be a fabrication technique. In welding, two or more metal parts are joined to form a single piece when one-part fabrication is expensive or inconvenient. Both similar and dissimilar metals may be welded. A variety of welding methods exist, including arc and gas welding, as well as brazing and soldering.

During arc and gas welding, the workpieces to be joined and the filler material (i.e., welding rod) are heated to a sufficiently high temperature to cause both to melt; upon solidification, the filler material forms a fusion joint between the workpieces.

Thus, there is a region adjacent to the weld that may have experienced microstructural and property

第十五章 金属工艺学

alterations; this region is termed the heat-affected zone (sometimes abbreviated HAZ).

焊接是一种不可拆卸的连接方法。在某种意义上，焊接可被认为是一种制造技术。如果一个零件完全用一种材料制造成本过高或者加工困难的话，可以用焊接将两个或更多的金属部分合并在一起。两个相似或不相似的金属件都可以焊接在一起。焊接包括电焊和气焊，以及钎焊和软钎焊等。

在电焊和气焊中，工件和填充材料（即，焊条）被加热到熔化；在钎焊加工中，填充材料熔化将工件连接在一起。

因此，毗邻焊缝的区域显微组织和属性可能已经发生了变化；这个区域被称为热影响区（有时缩写为HAZ）。

第五节 切削加工
Section 5 Cutting Processing

★ 主运动　primary motion
★ 进给运动　feed motion
★ 切削用量　cutting condition/cutting parameter
★ 切削速度　cutting speed
★ 进给量　feed rate
★ 背吃刀量（切削深度）　depth of cut
切屑厚度　chip thickness
切屑宽度　chip width
切削面积　area of cut
切削力　cutting force
切削热　heat in metal cutting
切削液　cutting fluid
金属切削机床　metal-cutting machine tool
★ 车床　lathe/turning machine
★ 钻床　drilling machine
台式钻床　bench-type drilling machine
立式钻床　vertical drilling machine
摇臂钻床　radial drilling machine
★ 镗床　boring machine
★ 刨床　planing machine
牛头刨床　shaping machine
龙门刨床　double column planing machine
插床　slotting machine/vertical shaping machine
★ 铣床　bed type milling machine
★ 磨床　grinding machine
外圆磨床　external cylindrical grinding machine
内圆磨床　internal cylindrical grinding machine
无心磨床　centerless grinding machine
平面磨床　surface grinding machine
滚齿机　gear hobbing machine
剃齿机　gear shaving machine
插齿机　gear shaping machine
磨齿机　gear grinding machine

Conventional machining is a form of subtractive manufacturing, in which a collection of material-working processes utilizing power-driven machine tools, such as saws, lathes, milling machines, and drill presses, are used with a sharp cutting tool to physically remove material to achieve a desired geometry. Machining is a part of the manufacture of many metal products, but it can also be used on materials such as wood, plastic, ceramic, and composites. Much of modern day machining is carried out by computer numerical control (CNC) machine. Computers are used to control the movement and operation of mills, lathes, and variety of other cutting machines.

Metal-cutting processing is a necessary instrument to obtain geometry form, dimension tolerance and surface mass of the part.

传统的切削加工是一种减法的制造，使用电动工具，如锯床、车床、铣床、钻床上锋利的刀具去除材料，获得所需的几何形状的零件。切削加工是金属产品制造中的一个重要方法，但是也可以用于其他材料，如木材、塑料、陶瓷和复合材料等。许多现代的加工都采用了计算机数控（CNC）机床。计算机用来控制铣床、车床和各种其他切削机床的运动和操作。

金属切削加工是获得零件几何形状、尺寸公差和表面质量的必要手段。

第十六章　机械设计课程设计
Chapter 16　Course Design of Machine Design

课程设计　course design
减速器　speed reducer
单级圆柱齿轮减速器　single helical gear speed reducer
齿数比　gear ratio
传动误差　transmission error
传动精度　transmission accuracy
范成法　generating method
齿轮形插刀　gear-form generating cutter
齿轮范成原理　principle of gear generating
齿条形插刀　rack-form generating cutter
蜗轮滚刀　worm gear hob
仿形法　copying method
轴肩挡圈　ring for shoulder
传动装置　transmission/driving
传动比　speed ratio/transmission ratio
增速比　speed increasing ratio
减速比　speed reducing ratio
无级变速　infinitely variable speed
摩擦轮传动　friction wheel drive
轴套　bearing bush
机械装置　mechanical device
机械动力装置　mechanical power device
机械效率　mechanical efficiency/mechanical advantage
内燃机　internal-combustion engine
简写符号　abbreviation
装配　assemble
齿轮传动装置　gear drive
布局　arrangement
增速装置　speed-increasing unit
减速装置　speed-decreasing unit
结构布置　physical arrangement
拟定技术条件　specify techniques
机械系统　mechanical system
原动机　prime mover
自然环境　physical environment
箱盖　housing (casing) cover
弹簧垫圈　spring washer
圆锥销　tapered pin
油标　oil gauge (gage)
放油塞　oil drain plug
纤维纸板垫片　fiber sheet
轴端挡圈　end check ring
轴　shaft
透端盖　bored end cover
斜齿大齿轮　helical-toothed wheel
减速器箱座　reducer housing (casing)
调整垫片　adjusting shim (liner)
密封圈　felt seal ring
齿轮轴　pinion shaft
起盖螺钉　depreasing bolt
排气装置　air exhauster device
观察孔盖　inspection hole cover

1 Mechanical design means the design of things and systems of a mechanical — machines, products, structures, devices, and instruments. For the most part mechanical design utilizes mathematics, the materials sciences, and the engineering-mechanics sciences.
机械设计是指机械装置和机械系统——机器、产品、结构、设备和仪器的设计。大部分机械设计需要利用数学、材料科学和工程力学知识。

2 There are different levels of design. When you design a machine which uses similar models, then you can design by imitation. Take one similar machine as a model, then, by keeping the main structure unchanged but changing some of the dimensions or sizes of the machine or replacing some parts with new ones, you can carry out design quickly. Such a design is called routine design. If you design a totally new machine or apply a new working principle in a machine, you have to create a new structure, not just imitate the existing one. This is creative design. Of course, creative design is more difficult than the routine one. Creative designs play an important part in developing new products to meet the growing demands of customers.
设计有不同的层次。如果采用类似的原型来设计机器,你可以通过模仿来设计。取一个相似的机器作为模型,然后保持主要结构不变,而改变机器的某些尺寸或大小,或者用新部件来替换其中的某些部件,这样你可以很快地完成设计。这样的设计称为常规设计。如果你要设计一个全新的机器,或在机器中应用新的工作原理,你就要创建一个新的结构,而不能仅仅模仿现有的。这就是创新设计。当然,创新设计要比常规设计难得多。在开发新产品以满足顾客逐渐增长的需求方面,创新设计起着很重要的作用。

3 Each gear has and serves its own particular application. There are several kinds of gears used in modern machinery. Some of these are spur, internal spur, helical, gear racks, bevel, worm and worm wheel.

每种齿轮都有适用于自己的特定应用领域。在现代机械中使用的齿轮有很多种,其中有直齿圆柱齿轮、内啮合直齿圆柱齿轮、斜齿轮、齿条、锥齿轮、蜗轮蜗杆等。

4 Speed reducer: the function of gearbox is to transmit rotational motion from a driving prime mover to a driven machine. The driving and driven equipment may operate at different speeds, requiring a speed—increasing or speed—decreasing unit. The gearbox therefore allows both machines to operate at their most efficient speeds. Gearboxes are also used to change the sense of rotation or bridge an angle between driving and driven machinery.

The gearbox configuration chosen for a given application is most strongly influenced by three parameters: physical arrangement of the machinery; ratio required between input and output speeds; torque loading (combination of horsepower and speed).

Other factors that must be considered when specifying a gear drive are: efficiency; space and weight limitations; physical environment.

减速器:齿轮箱的功用是将一种原动机的回转运动传递到从动机械上去。有些主、从动设备可能要按不同的速度运行,它们就需要增速装置或减速装置。齿轮箱应使主、从动设备能以各自最有效的速度运行。齿轮箱也用来改变回转方向,或者使主、从动机械之间呈某一角度。

下列三种参数最能影响选作一定用途的齿轮箱的结构:机械的结构布置;输入与输出之间要求的速比;扭矩载荷(马力和速度的组合)。

在具体选择一种齿轮箱时,还必须考虑其他方面的因素:效率;空间位置和重量的限制;自然环境。

5 Physical arrangement: the location of the driving and driven equipment in the mechanical system defines the input and output shaft geometrical relationship. Shaft arrangements can be parallel offset, concentric, right angle, or skewed. The material presented in this book focuses on parallel offset and concentric designs.

In the majority of parallel offset units in use, the input and output shafts are horizontally offset; however, vertical offsets are used and any orientation of input to output shaft is possible.

结构布置:主、从动设备在整个机械系统中所占的位置,确定了输入、输出轴之间几何位置的关系。轴的配置可以是平行轴偏置的、同轴式的、正交式的,也可以是斜交。本书所提供的资料集中阐明了平行轴偏置和同轴的设计。

在目前所使用的大多数平行轴偏置的装置中,输入、输出轴都是水平偏置的,然而也有垂直偏置的,其实从输入到输出轴之间呈任何一种方向都是允许的。

第十七章 控制工程
Chapter 17　Control Engineering

第一节　绪论
Section 1　Introduction

百科全书	encyclopedia	闭环控制	closed-loop control
使……自动化	automate	开环控制	open-loop control
离散时间系统	discrete-time system	传递函数	transfer function
阻断	interruption	相平面方法	phase-plane method
时间域技术	time-domain technique	现代控制理论	modern control theory
先验知识	priori knowledge	最优控制	optimal control
线性向量空间	linear vector space	低阶系统	low-order system
不确定性	uncertainty	非线性的	nonlinear
状态变量方法	state variable method	联立方程	simultaneous equations
动态特性	dynamic behavior	时变参数	time-varying parameter
常系数系统	constant coefficient system		

1 The idea of providing different signal paths for the process output and the command signal is a good way to separate command signal response from the response to disturbances.
对过程输出和指令信号提供不同的信号通路的想法是一个好办法,因为它能把指令信号响应与扰动响应相分离。

2 In this case, the integral action is realized as positive feedback around a first-order system.
在这种情况下是环绕一个一阶系统的正反馈实现积分作用的。

第二节　系统的数学模型
Section 2　The mathematical model of the system

解析模型	analytical model	偏微分方程	partial differential equation
随机系统	stochastic system	因变量	dependent variable
经验模型	experimental model	差分方程	difference equation
齐次的	homogeneous	传递函数矩阵	transfer function matrix
连续时间系统	continuous-time system	分段集中参数系统	lumped-parameter system
非齐次的	nonhomogeneous	拉普拉斯变换	Laplace transform
分布参数系统	distributed parameter system	概率统计的	probabilistic
输入输出方程	input-output equation	Z 变换	Z-transform
常微分方程	ordinary differential equation	梗概	bare-bones
独立变量	independent variable		

1 The input-output relationship in the Laplace transform domain is called the transfer function.
传递函数被定义为系统在拉普拉斯变换域中的输入输出关系。

2 The roots of the characteristic equation determine the stability of the system and the general nature of the transient response to any input.
特征方程的根决定系统的稳定性以及系统对任意输入的瞬态响应特性。

3 The second shifting theorem is useful in transforming delayed inputs and signals as transport lags.
第二平移定理适用于有输入延时、传输滞后的系统。

第三节 系统的时间响应分析
Section 3　The system analysis of time response

稳定性	stability	当且仅当	if and only if
时间响应	time response	一阶系统	first-order system
离散系统	discrete-time system	瞬态解	forced solution
干扰	disturbance	时间常数	time constant
稳定系统	stable system	初值	initial value
处于平衡状态	in equilibrium	二次延迟	quadratic lag
瞬态	transient phase	二阶系统	second-order system
稳态响应	steady-state response	阻尼比	damping ratio
激励	excitation	过阻尼	overdamped
指数的	exponential	临界阻尼	critically damped
振荡	oscillation	欠阻尼	underdamped
振幅	amplitude	单位阶跃输入	unit step input
脉冲响应	impulse response	原点	origin
复频域	complex frequency domain	主极点	dominating pole
频率域	frequency domain	性能指标	performance criteria
时间域	time domain	调节时间	setting time
频率响应	frequency response	超调	overshoot
传输延迟	transport lag	峰值时间	peak time
共轭对	conjugate pairs	上升时间	rise time
微分方程	differential equation	充要条件	necessary and sufficient condition
互异根	distinct roots	劳斯稳定判据	Routh stability criterion
瞬态模式	transient mode	无阻尼自然频率	undamped natural frequency
S 平面	S-plane	有阻尼自然频率	damped natural frequency
赫尔维茨判据	Hurwitz criterion	特征根	characteristic root

1 For system stability, the system pole must lie in the left half of the S-plane.
为使系统稳定,系统的极点必须位于 S 平面的左半平面。

2 An acceptable system must in minimum satisfy the three basic rules of stability, accuracy, and a satisfactory transient response.
一个满意的系统至少应满足三个基本准则:稳定性、精确性和满意的瞬态响应。

3 A necessary and sufficient condition for the system to be stable is that the roots of the characteristic equation have negative real parts.
系统稳定的一个必要条件为:系统的特征根具有负实部。

第四节 系统的频率特性分析
Section 4　The system analysis of frequency response

根轨迹	root locus	零点	zero
不可分解的	undecomposable	极点	pole
高阶	high order	相角条件	phase angle condition
增益	gain	幅值条件	magnitude condition
稳定程度	degree of stability	增益参数	parameter of interest
复变量	complex variable	幅角	argument
复变函数	complex function	极坐标形式	polar form
共轭复数	complex conjugate	特征值	eigen value

渐近线　asymptote
分离点　breakaway point
质心,心形曲线　centroid
相交　intersect
入射角　arrival angle
出射角　departure angle
动态补偿　dynamic compensation
零、极点图　pole-zero pattern
主动的,有源的　active

被动的,无源的　passive
相位超前　phase-lead
相位滞后　phase-lag
虚线　dashed curves
约束条件　constraint
逆时针方向　counterclockwise direction
比例微分控制　proportion-differential(PD)
比例积分控制　proportion-integral(PI)

1 These effects increase in strength with decreasing distance.
随着到原点距离的减小,它们的作用强度会增加。

2 In feedback system, the ratio of the output to the input does not have an explicitly factored denominator.
在反馈系统中,输出对输入的比没有明确的分母因数。

3 The system is stable if the trajectories do not change much and the initial condition is changed by a small amount.
如果初始条件有微小的变化,而轨迹却改变不大,则系统是稳定的。

第五节　系统的稳定性
Section 5　The stability of the system

伯德图　Bode plot
奈奎斯特图　Nyquist diagram
极坐标图　polar plot
常对数　common logarithm
分贝　decibel
最小相位　minimum phase
半对数坐标　semilog paper
镜像　mirror image
转折频率　break frequency
角频率　corner frequency

一阶滞后　simple lag
二阶滞后　quadratic lag
偏差　deviation
相角裕度　phase margin
幅值裕度　gain margin
穿越频率　crossover frequency
性能指标　performance specifications
相对稳定性　relative stability
稳态精度　steady-state accuracy
频率传递函数　frequency transfer function

1 All system characteristi croots is positive real part.
系统的全部特征根都具有正实部。

2 To ensure a specified attenuation of noise components in the input above a certain frequency (M in the Bode plot) should be below a certain level.
为保证对输入中高于一定频率的噪声成分指定的衰减,(伯德图中 M)应低于某一水平。

第六节　系统的性能指标与校正
Section 6　The performance index and correct of the system

给定精度　specified accuracy
带宽　bandwidth
优点,指标,准则　merit
串联　cascade
斜率　slope
超前补偿　lead compensation
滞后补偿　lag compensation
积分器　integrator

微分器　differentiator
多变量系统　multivariable system
记忆功能　memorization
系数矩阵　coefficient matrix
伴随矩阵　companion matrix
列向量　column vector
标量　scalar
精度　accuracy

漂移,偏移,偏差　drift
n 维空间　n-dimensional space
系统综合　system synthesis
术语　nomenclature
用户界面　user interface
原理图　schematic diagram
控制算法　control algorithm
补偿器　compensator
定性分析　qualitatively

定量分析　quantitatively
向量　vector
最优控制　optimal control theory
状态变量　state variable
约束　restriction
调节器　governor
稳定性　stability
系数　coefficient

1 This is also true for the leads corresponding to the simple and quadratic lag below.
对应于后面的一次和二次超前环节也是这样。

2 Adjust is to add new link in the system and to improve the performance of the system.
校正就是在系统中增加新的环节,以改善系统性能的方法。

3 The criteria have been to eliminate impulse disturbances and to drive the states of the system to zero.
其性能准则负责消减脉冲扰动并驱动系统的状态到零。

第十八章　机电测试技术
Chapter 18　Electromechanical Measuring and Its Testing

第一节　绪论
Section 1　Introduction

- ★ 信号　signal
- ★ 信息　information
- 回程误差　return error
- ★ 整流　commutate
- 整流器　commutator
- ★ 检波　demodulation/detection
- 检波器　radiodetector
- 振幅　amplitude
- 频率　frequency
- 压电　piezoelectricity
- 等效电路　equivalent circuit
- 光电　photoelectricity
- 磁电的　magnetoelectric
- 二极管　diode
- 三极管　dynatron
- 光电池　photocell
- ★ 灵敏度　sensitivity
- ★ 精确度　definition/precision
- ★ 可靠性　credibility
- ★ 线性的　linear range/confine

1 The aim, task, method, and development of testing are introduced in this section. Relative knowledge helps students know the whole survey of testing, and master the study method and other knowledge in future study.
本章简单介绍机电测试技术的目的、任务、方法、发展概况及相关内容、相关知识，使同学们对测试技术有一个总体、概括的了解，对其相关内容有一定的认识，并知道以后的学习中如何把握学习内容及方法。

2 The relations between signal and information are expounded.
阐述信号和信息的关系。

3 The basic constitution of testing system is explained.
了解测试系统的基本组成。

第二节　信号及其描述
Section 2　Signal and Its Description

- 机电一体化　mechatronics
- ★ 数据采集　data acquisition
- 平稳时间（平均故障间隔时间）　mean-time-between-failure (MTBF)
- 测试　test and measurement
- 终端负载　terminator
- 静态（台架、捕获、截获）实验　captive test
- ★ 误差　error
- 平均误差　mean error
- 相对误差、比例误差　relative/proportional error
- ★ 系统误差　systematic error
- ★ 静态误差　static error
- 人工误差　human-caused/respective error
- 随机误差　random error
- ★ 传感器　sensor
- 振荡器（交流发电机）　alternator
- 半导体　semiconductor
- 模拟的　analogous
- 数字的　digital
- 电阻　resistance
- 电容　capacitance
- 电感　inductance
- 电源　power source/electrical source
- 电极　pole/electrode
- 变压器　transformer
- 放大器　amplifier/magnifier
- 执行机构　executive/executant/executor
- 反馈　feedback

第十八章 机电测试技术

1 The classification and relative examples are introduced in this section to help students master the basic method, aim, meaning and the express way of signal intention.
本章介绍信号的分类,并举例说明各种类型的信号,同时帮助学生掌握信号分析的基本方法、分析目的和意义,了解信号强度的表述方式。

2 To be skilled in the working principle of resistance, capacitance and inductance.
正确掌握电阻、电容、电感等电器元件的工作原理并熟练使用。

3 Human-caused error can be avoided, but systematic error can not be avoided.
人工误差可以避免,系统误差不可避免。

第三节 信号及其描述
Section 3 Signal and Its Description

★ 输入输出 input output	电荷 electric charge
确定的 assured	定理 theorem
平衡 balance/counterpoise/equilibrium	量化 quantity
规律 law/orderliness	编码 coding
规律的 regular	混叠 compound/mixed
周期 period/cycle	重叠 wrap/lap
★ 周期性 periodicity	窗函数 window function
离散的 disperse/scatter	矩形函数 rectangle function
频率 frequency	三角形函数 triangle function
★ 减振器 vibration isolator	自相关 self-correlation
★ 调制 modulate/confect	互相关 commutation correlation
★ 调制器 modulator	安培表 ampere meter
调制解调器 modem	微分放大器 differentiating amplifier
电流 current	扫描 sweep
电压 voltage	矢量分析 vector analysis
电容器 capacitor	模拟分析法 model analysis
直流 direct current	动态分析器 dynamic analyzer
交流 alternating current	顶角 apical angle
滤波器 filter	锐角 acute angle
低通 low-frequency	钝角 obtuse angle
载波 carrier wave	折光天线 dioptric antenna

1 The relation among testing, computation and measure is introduced to everyone. The basic character of testing device is emphasized to realize the composition and condition of precision. It is important for us to master and use the testing device.
介绍测试、计量、测量之间的关系,测试装置的组成,着重介绍测试装置的基本特性,这对我们掌握和使用测试装置有十分重要的作用。

2 AM, FM and PM are three different methods of signal modulation.
调幅、调频、调相是三种不同的信号调制方式。

3 The testing device is composed of input, output and exchange device.
测试装置包括输入、输出和中间变换装置。

第四节 传感器
Section 4 Sensor

减光器(减声器) diminisher	引出端 leading-out end
二地址 double-address	无定形扫描仪 espews

量化误差	quantization error	反射板	baffle board
奇偶校验	generator-checker	充电盘	charging board
激光扫平仪	geoplane	谱线展现	line broadening
光电传感头	photohead	扬声器	speaker
光电效应	photoeffect	多芯电缆	bank cable
光电二极管	photodiode	单芯电缆	monofiber-cable
敏感元件	sensitive cell	阴极脉冲调制的	cathode-pulsed
模拟记录器	instrumentation recorder	★测力计	ergometer
全波整流	full-wave rectification	★磁元件	magnetic cell
半波整流	single-wave rectification/one-half period rectification	热传导	heat conductor
		良导体	good conductor
差配继电器	differential relay	修正的,补偿的	correcting
双线性	bilinearity	余弦	cosine
二进制处理	bit-manipulation	反余弦	arc cosine
双变量的	bivariate	全微分	total differential
程序块	programme block	偏微分	partial differential
★数字块	digital block	★数字器	digitizer

1 Sensors are classified in this section, in the meanwhile, all kinds of sensors and their principles are introduced, for example, mechanical sensor, resistance sensor, electric capacity sensor, inductance sensor, piezoelectricity sensor, semiconductor sensor, and optical sensor.
本章主要介绍传感器的分类,同时介绍机械式传感器、电阻式传感器、电容式传感器、电感式传感器、压电式传感器、半导体式传感器、光学传感器等各种传感器的原理。

2 The sensor is composed of sensitive elements and auxiliary devices.
传感器是由敏感元件和辅助装置组成的。

3 The sensor is a device that is attached to something measured and transduced to the same or another signal.
传感器是直接作用于被测量体,并能按一定规律将其转换成同种或别种量值输出的器件。

第五节 信号调整、处理
Section 5　Modulating and Disposing Signal

频谱	frequency	计量	computation measure
振动拾取	vibration and shock pickup	量程	range
测试	calibration	噪音	noise
强度	intension/intensity/strength	积分	integral
峰值	peak value	微分	differential
均值	average	数字信号处理	digital signal processing, DSP
绝对均值	absolute average	随机信号储存器/只读存储器	RAM/ROM
有效	efficiency/availability	★激振器	oscillator
功率	power	★示波器	oscillograph
付立叶变换	fast fourier transform	★信号发生器	signal generator
级数	progression/series	导线	lead
奇数	odd number	实验者	experimenter
偶数	even number	实验室	laboratory
对称	symmetry	位移	displacement
样本	sample/specimen	速度	quickness

速率　rate
加速度　acceleration
温度　temperature
湿度　humidity
力矩　moment of force
转力矩　torque
元件　component/element
电路　circuit
噪声去除器　noise-abater
液压减振器　hydraulic absorber

1 The signal from sensor must be disposed to be the useful signal that we need. All methods of modulation and record are introduced to benefit analysis, for example, electric bridge, modulation and demodulation, the transformation of analogue and digital, filter, self-correlation, commutation correlation and so on.
通过传感器得到的信号需要进一步处理才能够成为我们所需要的可以分析的有用信号,本章介绍信号的调整方法、处理手段以及对信号如何记录等,如电桥、调制与解调、A/D变换、D/A变换、滤波、自相关、互相关等帮助我们得到有用的测试信号,便于分析。

2 Electric bridge is a kind of electric circuit composed of resistance, inductance or capacitance.
电桥是将电阻、电感或电容等参量的变化变为电压或电流输出的一种测量电路。

3 The modulation includes AM, FM and PM.
调制可分为调幅、调频、调相。

第六节　信号的记录
Section 6　Recording Signal

★阻尼器　damper
总衰减　total attenuation
★噪声衰减　noise attenuation
引力　attraction
噪声听度计　noise-audiometer
语言测听计　live-voice audiometer
★确定性/可靠性　authenticity
变容二极管　variable capacitance diode
设计自动化　design automation
数字式自动化　digital automation
自动取样器　autosampler
自动扫描　autoscan
★自谱　autospectrum
自激振动　self-in-duced vibration
★样本均值　sample average
★总体平均　overall average
加样平均值　weighted average
返程　back stroke
反向电阻　back-resistance
视频带　video band
音频带　voice band
抗体　antisubstance
测量孔径　measuring aperture
抗尺器　stretching apparatus
放影仪　schlieren apparatus
加力(加负荷)　stress application

演绎法　deductive approach
探试法　heuristic approach
最小二乘逼近　least square approximation
牛顿近似法　Newton's approximation
主轴　principle axis
集中/分散制优器　centralized/decentralized arbiter
微处理机制优器　microprocessor arbiter
绝缘材料　insulating material
数据范围　data area
定点运算　fixed-point arithmetic
浮点　floating-point
力偶　couple
力矩臂　moment arm
天平杆　scale arm
宏汇编程序　macroassembler
等离子体加速器　plasma accelerator
加速度计　accelerometer
控制存取　control access
随机存取　random access
存取地址　access-address
辅助设备　accessory
重复精度　repetitive accuracy
测量精度　measuring accuracy
旋转精度　running accuracy
扫描　sweep

1 Several experiments are also introduced in this section, including the composition of mechanical and electric testing, force measure and displacement measure.

本章将介绍几种实验,包括机电测试系统的组成、力测量、位移测量等实验。

2 The signal tested must be recorded and shown by some devices. Some kinds of recording and showing devices are introduced in this section, for example, magnetoelectric indication device, radial oscillograph, servo recorder and so on.

对采集到的信号,要通过一定的装置进行显示和记录,本章介绍几种显示和记录的装置,如动圈式磁电指示机构、光线示波器、伺服式记录仪等。

3 Darsonval mechanism is the core of pen register and beam oscilloscope.

达松瓦尔机构是笔式记录器和光线示波器的核心部分。

第二部分 应用篇

一、摘要写作
1. Abstract Writing

1 What are Research Abstracts?
An abstract is a stand-alone statement that briefly conveys the essential information of a paper and presents the objective, methods, results, and conclusions of a research project.
什么是研究摘要？
摘要是一个独立性的描述,简洁地传达了论文的重要信息,提出了研究项目的目的、方法、结果和结论。

2 Functions/Parts of Abstract 摘要的功能或组成
(1) background information 背景信息
A simple opening sentence or two can be used to make clear the work in context.
一或两个开场句子交代工作背景。
(2) purpose of the study/principal activity 研究目的或主要研究活动
One or two sentences can be used to provide the purpose of the work.
一或两个句子给出工作目的。
(3) method 研究方法
One or two sentences can be used to explain what problems have already been solved.
一或两个句子阐述解决的问题。
(4) research results 研究结果
One or two sentences can be used to indicate the main findings.
一或两个句子表明主要的结果。
(5) conclusion/recommendation 结论或建议
Use one sentence or two to state clearly the most important conclusion of the work.
一或两个句子表明研究工作的最重要的结论。

3 Sample 实例

(1)

Abstract: Edge CAD/CAM of milling cutter need edge checking to ensure the precision of milling cutter for screw rotor. The edge checking system makes use of CNC technology in the movement control of system to enhance the precision of edge checking. The calculation of theoretical tool path of detector and theoretical model of cutter edge were indicated. The paper discussed was made up of hardware. The software of edge checking system was programmed in VB. Main program structure was discussed. The dynamic link library (DLL) was founded for I/O channel operating. By using CNC technology, the measurement accuracy is increased on a large scale, and its resolution is within $0.5 \mu m$.

摘要:铣刀刃形曲线的加工精度决定了螺杆转子的加工精度,因此对刃形曲线的精确、高效检测是保证螺杆转子铣刀刃形 CAD/CAM 的重要手段之一。螺杆转子铣刀刃形 CAD/CAM 的刃形检测系统是采用 CNC 进行运动控制的刃形数据采集分析处理系统(DAS)。文章重点阐述了系统的硬件组成、理论刃形模型建立、测头/工作台运动控制轨迹的计算、自动编程以及运动控制原理。系统软件部分论述了 VB 主体程序的结构和建立用于 I/O 通道操作的动态链接库(DLL),以及通过调用 DLL 实现数据采集、运动控制的主要设计思想。利用这种 CNC 技术可以使系统测量精度大幅度提高，其分辨率可达 $0.5 \mu m$。

(2)

Abstract: A new method of pattern recognition of tool wear based on Discrete Hidden Markov Models (DHMM) is proposed to monitor tool wear and to predict tool failure in cutting process. At first FFT features are extracted from the vibration signal and cutting force in cutting process, then FFT vectors are presorted and coded into code book of integer numbers by SOM, and these code books are introduced to DHMM for machine learning to build up 3-HMMs for different tool wear stage. And then, pattern of HMM is recognized by using maximum probability. Finally the results of tool wear recognition and failure prediction experiments are presented and show that the method proposed is effective.

摘要：对于金属切削过程中的刀具磨损，提出了基于隐马尔可夫模型（DHMM）的模式识别理论，识别刀具不同磨损状态，从而预报刀具破损的新方法。该方法对切削过程中的切削力信号动态分量和刀柄振动信号进行 FFT 特征提取，然后利用自组织特征映射（SOM）对提取的特征矢量进行预分类编码，把矢量编码作为观测序列引入到 DHMM 中进行机器操作学习，建立三个不同磨损状态的 HMM 模型；并利用最大概率进行模式识别。实验表明，该方法对车刀磨损过程进行识别和预报是十分有效的。

(3)

Abstract: In envelope process of cutter, profile and tool-path of cutter are complicated, and the grinding depth is variable which causes grinding force to change on a large scale and the error of edge envelope to increase. To improve cutter accuracy, the theoretical model of cutter edge and theoretical tool-path of CNC tool grinder were calculated. Closed-loop control system was designed to accomplish control of envelope process. By adjusting feed-rate, constant force grinding process is accomplished to improve the precision of edge. The methods of edge design, tool-path calculation and manufacture processing of cutter on CNC tool grinder were discussed. Machining test indicates that cutter error is $8 \sim 11 \mu m$, and coherence error of sample thickness is reduced to only $3 \mu m$, by constant force grinding.

摘要：铣刀刃形包络中，由于其结构和加工时刀具轨迹的复杂性，磨削深度变化很大，因此会引起磨削力的大范围变化从而引起铣刀刃形的加工精度下降。为了提高铣刀的包络精度，利用刃形的理论模型计算 CNC 工具磨床的修磨轨迹，并设计了刃形包络的闭环控制系统，完成刃形包络成型过程。在包络过程中通过改变进给速率实现恒力磨削进而提高了刀具的加工精度。文章讨论了刃形设计，修磨轨迹计算和刀具在 CNC 工具磨床的磨削工艺。加工实验表明：刀具的刃形误差为 $8 \sim 11 \mu m$，由磨削力引起的复映误差仅为 $3 \mu m$。

(4)

Abstract: Empirical Mode Decomposition (EMD) is an adaptive decomposition method developed recently in non-stationary signal processing field. But one of the major drawbacks of the EMD is that, when two individual components in a signal with frequencies within an octave usually can't be decomposed by normal EMD method. The paper presents a new modified method based on Frequency Heterodyne EMD which was used to solve the drawback of normal EMD. The simulation and application in backlash nonlinearity system show that the proposed method can improve the resolving power of EMD with high accuracy and easy operation, and can also successfully separate steady component and transient component from the output, which provide a base for deep research of backlash nonlinearity systems. And it also validates the feasibility of the new method.

摘要：经验模式分解（EMD）是近年来非平稳信号处理领域出现的一种自适应分解方法。EMD 存在的一个主要缺点是，当信号中两组成分量的频率在二倍频内时，EMD 无法将两者分解开。为了提高 EMD 的频率分辨率，本文根据频率外差 EMD 方法提出了一种改进算法。仿真分析和间隙非线性系统的应用表明，该算法能有效提高 EMD 的频率分辨率，其分解精度高，操作简单，且能成功分离系统响应输出的稳态信息和暂态信息，为间隙非线性系统的深入研究提供了基础，同时也证明了该方法的可行性。

二、实验设备介绍
2. Experiment Equipment Introduction

Lathe 车床

The purpose of a lathe is to rotate a part against a tool whose position it controls. It is useful for fabricating parts and/or features that have a circular cross section. The spindle is the part of the lathe that rotates. Various workholding attachments such as three jaw chucks, collets, and centers can be held in the spindle. The spindle is driven by an electric motor through a system of belt drives and/or gear trains. Spindle speed is controlled by varying the geometry of the drive train.

The tailstock can be used to support the end of the workpiece with a center, or to hold tools for drilling, reaming, threading, or cutting tapers. It can be adjusted in position along the ways to accomodate different length workpieces. The ram can be fed along the axis of rotation with the tailstock handwheel.

The carriage controls and supports the cutting tool. It consists of:
A saddle that mates with and slides along the ways;
An apron that controls the feed mechanisms;
A cross slide that controls transverse motion of the tool (toward or away from the operator);
A tool compound that adjusts to permit angular tool movement;
A toolpost T-slot that holds the toolpost.

车床的用途是将旋转的工件向刀具进给。它非常适合加工回转体。车床上的主轴也一起旋转。主轴可以安装不同的夹紧附件,如三爪夹盘、弹性夹头和顶尖。主轴通过电机经皮带和齿轮传动来驱动。主轴转速通过改变轮系的传动比来控制。

使用尾座上的顶尖可以支承工件的尾部,尾座还可以用来夹持刀具进行钻孔、铰孔、攻螺纹及加工锥孔。尾座可以调整在导轨上的位置以适应不同长度的工件。使用尾座上的手轮可以让套筒沿着其旋转轴线方向进给。

大溜板控制和支承刀具。它的组成如下:
与导轨配合的溜板座;
控制进给机构的溜板箱;
用于控制刀具横向运动(向着或远离操作者)的横向溜板;
用于调整刀具角度的刀座;
用于夹持刀座的T形槽。

Flexible Manufacturing System 柔性制造系统

There is a TVT-4000E Flexible Manufacturing System in the lab, indicated as Fig 1.
实验室有一 TVT-4000E 柔性制造系统,如图1所示。

Five-axis machining center 五轴加工中心

Five-axis linkage processing center is a kind of lathe with high scientific and technological content, high precision, specially designed for processing complex curved surfaces. This kind of lathe has significant impact on the aviation, aerospace industry, military, scientific research, precision instruments, high precision medical equipment industry and so many other aspects of a nation. Currently, it is generally believed that the five-axis linkage numerical control machine tool system is the unique method to process the linkage of impeller, leaf, marine propeller, heavy generator rotor, steam turbine rotor, and large diesel engine crankshaft. Thus, it is prohibited to export high precision five-axis

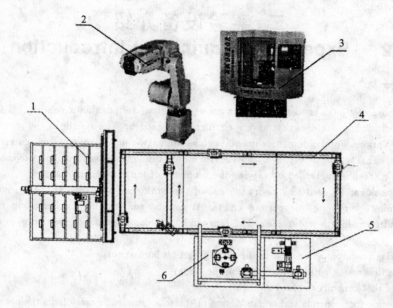

图 1 柔性制造系统
Fig. 1 Flexible Manufacturing System
1—立体仓库系统单元（Three-dimensional warehouse system unit） 2—FANUC 机器人单元（FANUC robot unit） 3—数控铣床（numerical control milling machine） 4—多工位环行线系统单元（multi-position progressive system unit） 5—供给系统单元（Supply system unit） 6—视觉五维装配系统单元（Visual five-dimensional assembly system unit）

machine tools in the US, European union and Japan.

There exists a DMU 60 monoBLOCK five-axis machining center in the lab, it is a machine with excellent dynamic performance, a spindle of quick liner motion and rotating, and a shorter time of tool changing, such properties facilitate the producing efficiency greatly. B axis swing angle is up to 150°, besides, DMU 60 monoBLOCK processing center fulfils the need of teaching and producing with real-time collision monitoring and other optional accessories. Unlimited power, maximum flexibility: DMU monoBLOCK adopts modular structure design, and available rotate speed is from 10 000 rpm to 42 000 rpm.

五轴联动加工中心是一种科技含量高、精密度高并专门用于加工复杂曲面的机床,这种机床系统对一个国家的航空、航天、军事、科研、精密器械、高精医疗设备等行业有着举足轻重的影响力。现在,大家普遍认为,五轴联动数控机床系统是解决叶轮、叶片、船用螺旋桨、重型发电机转子、汽轮机转子、大型柴油机曲轴等加工问题的唯一手段。也是鉴于此,美国、欧盟及日本一直将一些高精度的五轴机床列为限制出口产品。

实验室拥有一台德玛吉五轴联动 DMU 60 monoBLOCK 加工中心,它具有极好的动态性能、快速直线和回转轴以及更短的换刀时间,这些性能极大地缩短了加工时间。B 轴摆动角度达 150°,另外 DMU 60 monoBLOCK 加工中心还具有实时碰撞监控和丰富的选配件,极大地满足了教学和生产的需要。功率无极限,柔性最大：DMU monoBLOCK 机床采用模块化结构设计,可选配转速从 10 000 rpm 到 42 000 rpm 的不同电主轴。

二、实验设备介绍 117

图2 DMU 60 monoBLOCK 加工中心
Fig. 2 machining center

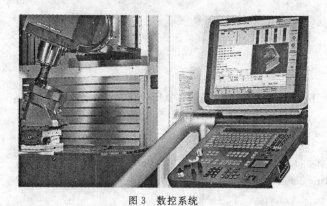

图3 数控系统
Fig. 3 Numerical Control System

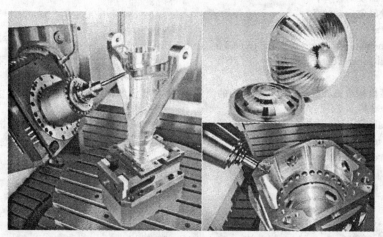

图4 应用案例
Fig. 4 Applications

三、大连民族学院课程实验报告单
3. Experimental Report of Dalian Nationalities University

（课程名称：机械原理实验）
(Course name：Experiment Report of Theory of Machines and Mechanisms)

Professional Classes：＿＿＿＿＿＿＿＿＿＿
专业班级：＿＿＿＿＿＿＿＿＿＿＿＿＿＿
Name：
姓名：
Student ID：
学号：
Date：
实验日期：
Score：
成绩：
Instructor：
指导教师：

The Experimental Name：Map of Kinematic Diagram
实验名称：机构运动简图测绘

1 Experiment Purpose：
Grasp how to map the Kinematic Diagram;
Consolidate the calculation of the degree of freedom;
Verify the kinematic determination.
实验目的：
掌握机构运动简图测绘的基本方法；
巩固机构自由度的计算；
验证机构做确定运动的条件。

2 Experimental Content
Map the kinematic diagrams of some typical institutions or machines; calculate the degree of freedom; verify the kinematic determination.
实验内容
对一些典型机构或机器进行机构运动简图的测绘，计算其自由度，并验证其运动是否确定。

3 Experimental Equipment
Some typical institutions or machines;
Measuring tool;
Pencil, eraser, white paper, set square and compasses.
实验用具
典型机构或机器若干件；
量具；
铅笔、橡皮、白纸、三角板和圆规。

4 Theory of Mapping
From the viewpoint of the movement, the mechanism movement is only related with the number of members in the mechanism, the type and number of the pairs, and relative position between the

pairs, regardless of the complex shape of members and the specific structure of pairs. Therefore, when conducting institutional analysis, the factors that have nothing to do with mechanism movement are often not considered. Only simple lines and the predetermined symbols are used to represent the pairs and the members, and the relative position between the pairs is mapped in accordance with a certain proportion. The correct kinematic simple graph is called kinematic diagrams.

测绘原理

从运动的观点来看,机构的运动仅与机构中构件的数目,各构件组成的运动副的类型、数目以及各运动副之间的相对位置有关,而与构件的复杂外形和运动副的具体结构无关。因此,在进行机构分析时,常常不考虑那些与机构运动无关的因素,而仅用简单线条和规定符号来表示运动副和构件,并按一定比例表示各运动副间的相对位置。这种能正确表达机构运动特性的简单图形称为机构运动简图。

5 Experimental Procedure

实验步骤

Map the kinematic diagrams with pencils.

用铅笔绘出机构运动简图。

Calculate the degree of freedom of the mechanism and verify the kinematic determination.

计算所测绘机构的自由度,并验证此机构是否具有确定的运动。

6 Data

实验数据

Name of the mechanism: slider-crank mechanism

机构名称:曲柄滑块机构

Kinematic diagram:

机构运动简图:

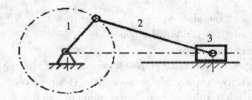

Scale: $\mu_l = 1m/10mm$

比例尺: $\mu_l = 1m/10mm$

Calculate the degree of freedom: $F = 3n - 2P_l - P_h = 3 \times 3 - 2 \times 4 - 0 = 1$

计算自由度: $F = 3n - 2P_l - P_h = 3 \times 3 - 2 \times 4 - 0 = 1$

The kinematic determination of this mechanism is verified.

该机构具有确定的运动。

7 Questions

What can we see from the kinematic diagram?

The number of members in the mechanism, the type and number of the pairs, and relative position between the pairs.

思考题

机构运动简图能反映实际机构的哪些内容?

机构中构件的数目,各构件组成的运动副的类型、数目以及各运动副之间的相对位置。

四、常用软件介绍
4. The Introduction of Common Software

(1) Introduction of MATLAB

MATLAB is a powerful computing system for handling the calculations involved in scientific and engineering problems. The name MATLAB stands for MATrix LABoratory, because the system was designed to make matrix computations particularly easy. MATLAB was originally written to provide easy access to matrix software developed by the LINPACK and EISPACK projects. Today, MATLAB engines incorporate the LAPACK and BLAS libraries, embedding the state of the art in software for matrix computation.

MATLAB has evolved over a period of years with input from many users. In university environments, it is the standard instructional tool for introductory and advanced courses in mathematics, engineering, and science. In industry, MATLAB is the tool of choice for high-productivity research, development, and analysis.

MATLAB is a programming environment for algorithm development, data analysis, visualization, and numerical computation. As a high-performance language for technical computing, it can solve technical computing problems faster than with traditional programming languages, such as C, C++, and Fortran.

Typical uses of MATLAB include
- Math and computation
- Algorithm development
- Data acquisition
- Modeling, simulation, and prototyping
- Data analysis, exploration, and visualization
- Scientific and engineering graphics
- Application development, including graphical user interface building

MATLAB is an interactive system whose basic data element is an array that does not require dimensioning. This allows you to solve many technical computing problems, especially those with matrix and vector formulations. It would take to write a program in a scalar noninteractive language such as C or Fortran in a fraction of the time.

Toolboxes

MATLAB features a family of add-on application-specific solutions called toolboxes. Very important to most users of MATLAB, toolboxes allow you to learn and apply specialized technology. Toolboxes are comprehensive collections of MATLAB functions (M-files) that extend the MATLAB environment to solve particular classes of problems. Areas in which toolboxes are available include signal processing, control systems, neural networks, fuzzy logic, wavelets, simulation, and so on.

The MATLAB System

The MATLAB system consists of five main parts: development environment, the MATLAB mathematical function library, the MATLAB language, the MATLAB graphics, the MATLAB application program interface (API).

矩阵实验室(MATLAB)简介

矩阵实验室(MATLAB)是一个解决科学和工程计算等问题的强大的计算系统。MATLAB 的名称源于矩阵实验室的缩写,因为该软件使复杂的矩阵计算变得非常容易。MATLAB 软件最初是为了方便进入由 LINPACK 和 EISPACK 项目开发的矩阵软件而开发的。今天,MATLAB 中合并了 LAPACK 和 BLAS 两个库,嵌入到 MATLAB 软件用于矩阵运算。

MATLAB 历经多年的发展演变,吸引了越来越多遍及各领域的使用者。在学校里,它是一个标准的初、高级课程的教学工具,应用于数学、工程和科学研究的各个方面。在工业生产中,MATLAB 也是一个用于高生产率的研究、开发和分析的工具。

MATLAB 拥有用于算法开发、数据分析、可视化和数值计算的程序设计环境。作为一种用于专业技术计算的高级语言,MATLAB 比传统的 C、C++ 和 Fortran 有着更快的计算速度。

MATLAB 典型的应用包括:
- 数学和计算
- 算法开发
- 数据采集
- 建模、仿真和系统原型的构建
- 数据分析,研究和可视化
- 科学和工程图形的绘制
- 应用程序的开发,包括图形用户口界面的创建

MATLAB 是一个交互式系统,其基本数据元素是一个不需要标注数组。这可以解决很多专业技术计算问题,尤其是那些矩阵和向量表达式。它可在短时间内,像 C 或 Fortran 一样,用标量的非交互式语言编写程序。

工具箱

MATLAB 的显著特点是它的工具箱。这些工具箱把各种特定应用领域解决方案组合在一起。这对大多数 MATLAB 用户来说是非常重要的。工具箱可以让你学习和运用某一领域的特殊专业技术。工具箱也是由许多 MATLAB 函数组成的集合。这些 MATLAB 函数用于解决 MATLAB 环境下的各种专业问题。全面 MATLAB 函数(M 文件)扩展了 MATLAB 环境,以解决问题的特定类别。工具箱所涉及的常见领域包括信号处理、控制系统、神经网络、模糊逻辑、小波分析、仿真等等。

MATLAB 系统

MATLAB 系统有 5 个主要组成部分,分别为 MATLAB 的开发环境、MATLAB 的数学函数库、MATLAB 语言、MATLAB 图形、MATLAB 的应用程序界面(API)。

(2) Introduction of Mathematica

Mathematica is a scientific calculation software, and is a good combination of numerical and symbolic computation engine, graphics systems, programming languages, text system, and with other application advanced connection. Many functions in the corresponding field are in a leading position in the world. As of 2009, it is now one of the most widely used mathematical software. The release of Mathematica marks the beginning of modern science and technology calculation. Since the nineteen sixties, in numerical, algebraic, graphical, and other aspects of a wide range of applications, Mathematica is in the most powerful system amone the world general purpose computing systems. Since 1988 been released in 1988, it has a profound impact on how to use computer in technology and other areas.

Mathematica 简介

Mathematica 是一款科学计算软件,很好地结合了数值和符号计算引擎、图形系统、编程语言、文本系统和其他应用程序。很多功能在相应领域内处于世界领先地位。截至 2009 年,它也是使用最广泛的数学软件之一。Mathematica 的发布标志着现代科技计算的开始。自 20 世纪 60 年代以来,在数值、代数、图形和其他方面应用广泛,Mathematica 是世界上通用计算系统中最强大的系统。自 1988 发布以来,它已经对如何在科技和其他领域运用计算机产生了深刻的影响。

(3) Introduction of MathCAD

MathCAD is an engineering calculation software of the United States of America's PTC company. As the global standard for engineering calculation, and proprietary computational tools and electronic form, MathCAD allows engineers to use the detailed application of mathematical functions and dynamic, perceivable unit calculation to design and recording engineering calculation at the same time. Unique visual format and scratch pad interface intuitive, standard mathematical symbols, text and graphics were integrated into a work table. MathCAD uses to write on the blackboard formula way to allow users to express the problem, through the computation of the underlying engine calculation results returned and displayed on the screen. Calculation of approximate transparent process enables the user to focus on the reflection of the problem rather than the tedious step.

MathCAD 简介

MathCAD 是美国 PTC 公司旗下的一款工程计算软件，作为工程计算的全球标准。它与专有的计算工具和电子表格不同，MathCAD 允许工程师利用详尽的应用数学函数和动态、可感知单位的计算来同时设计和记录工程计算。独特的可视化格式和便笺式界面将直观、标准的数学符号、文本和图形均集成到一个工作表中。MathCAD 采用类似在黑板上写公式的方式让用户表述所要求解的问题，通过底层计算引擎计算返回结果并显示在屏幕上。计算过程近似透明，使用户专注于对问题的思考而不是繁琐的求解步骤。

(4) Introduction of Maple

Maple is the world's most commonly used mathematical and engineering calculation software, in math and science are "reputation, mathematician known software". Maple in the world with millions of users, is widely applied in the fields of science, engineering and education and other fields, with the user penetration over 96% of the world's major universities and research institutes, over 81% of the fortune in five hundred companies.

Maple system modeling and simulation of built-in advanced technology solve the mathematics problems, including the world's most powerful symbolic computation, infinite precision numerical calculation, innovative internet connection, powerful 4GL language, built more than 5000 calculation command, mathematical and analytical functions covering almost all branches of mathematics, such as calculus, differential equations, special functions, linear algebra, image and sound processing, statistics, power system

Maple not only provides programming tools, more important is to provide knowledge of mathematics. Maple is the essential scientific computing tools for professor, researchers, scientists, engineers, students, from simple digital computation to highly complex nonlinear problems, Maple can help you quickly, efficiently to solve the problem. Users through the Maple products can achieve areas of physical system modeling and simulation in a single environment, symbolic computation, numerical calculation, program design, technical documents, presentations, algorithm development, external program to connect and other functions, to meet the different levels of the needs of users, from high school students to senior researchers.

Maple 简介

Maple 是目前世界上最为通用的数学和工程计算软件之一，在数学和科学领域享有盛誉，有"数学家的软件"之称。Maple 在全球拥有数百万用户，被广泛地应用于科学、工程和教育等领域，用户渗透超过 96% 的世界主要高校和研究所，超过 81% 的世界财富五百强企业。

Maple 系统内置高级技术解决建模和仿真中的数学问题，包括世界上最强大的符号计算、无限精度数值计算、创新的互联网连接、强大的 4GL 语言等，内置超过 5000 个计算命令，数学和分析功能覆盖几乎所有的数学分支，如微积分、微分方程、特殊函数、线性代数、图像声音处理、统计、动力系统等。

  Maple 不仅仅提供编程工具，更重要的是提供数学知识。Maple 是教授、研究员、科学家、工程师、学生们必备的科学计算工具，从简单的数字计算到高度复杂的非线性问题，Maple 都可以帮助您快速、高效地解决问题。用户通过 Maple 产品可以在单一的环境中完成多领域物理系统建模和仿真、符号计算、数值计算、程序设计、技术文件、报告演示、算法开发、外部程序连接等功能，满足从高中学生到高级研究人员各个层次用户的需要。

(5) Introduction of ANSYS

  ANSYS software is the large-scale general finite element analysis software blended with structure, fluid, electric field, magnetic field, sound field analysis. Developed by the world's largest finite element analysis software company of the United States of America ANSYS, it can realize data sharing and exchange with most CAD software interface, such as Pro/Engineer, NASTRAN, Alogor, AutoCAD, I-DEAS, is one of senior CAE tools in modern product design.

ANSYS 简介

  ANSYS 软件是融结构、流体、电场、磁场、声场分析于一体的大型通用有限元分析软件。它由世界上最大的有限元分析软件公司之一的美国 ANSYS 开发，能与多数 CAD 软件接口，实现数据的共享和交换，如 Pro/Engineer、NASTRAN、Alogor、AutoCAD、I-DEAS 等，是现代产品设计中的高级 CAE 工具之一。

(6) Introduction of ADAMS

  ADAMS, automatic dynamic analysis of mechanical systems (Automatic Dynamic Analysis of Mechanical Systems), the software is virtual prototype analysis software developed by the United States MDI (Mechanical Dynamics Inc.) At present, ADAMS has been adopted by hundreds of main manufacturers of all walks of life in the world. According to the 1999 dynamic simulation analysis of mechanical system software international market share statistics, ADAMS software sales totaled nearly eighty million, held the share of 51%, now have been incorporated into the American MSC company.

ADAMS 简介

  ADAMS，即机械系统动力学自动分析（Automatic Dynamic Analysis of Mechanical Systems），该软件是美国 MDI 公司（Mechanical Dynamics Inc.）开发的虚拟样机分析软件。目前，ADAMS 已经被全世界各行各业的数百家主要制造商采用。根据 1999 年机械系统动态仿真分析软件国际市场份额的统计资料，ADAMS 软件销售总额近八千万美元，占据了 51% 的份额，现已经并入美国 MSC 公司。

(7) Introduction of Moldflow

  Autodesk Moldflow simulation software with injection molding simulation tool can help you to validate and optimize plastic parts, injection mold and injection molding process. The software is capable to provide guidance for design, mold production staff, engineers, through simulation settings and results elucidates to display wall thickness, gate location, material, and geometry changes to affect manufacturability. From the thin wall parts to thick, solid parts, Autodesk Moldflow geometric graphics support can help users in the final design decision before the test scenarios.

Moldflow 简介

  Autodesk Moldflow 仿真软件具有注塑成型仿真工具，能够帮助您验证和优化塑料零件、注塑模具和注塑成型流程。该软件能够为设计人员、模具制作人员、工程师提供指导，通过仿真设置和结果阐明来展示壁厚、浇口位置、材料、几何形状变化如何影响可制造性。从薄壁零件到厚壁、实体零件，Autodesk Moldflow 的几何图形支持可以帮助用户在最终设计决策前试验假定方案。

(8) Introduction of SolidWorks

SolidWorks 3D CAD software offers three packages building in functionality which are tiered to best suit the needs of your organization. All packages utilize the intuitive SolidWorks user interface to speed your design process and make you instantly productive.

SolidWorks 简介

SolidWorks 3D CAD 软件提供了三种内置功能的软件包，这些软件包分成不同的层次，您的组织可使用最符合需要的软件包。所有软件包都使用直观的 SolidWorks 用户界面来加快您的设计过程并可立即提高您的生产效率。

SolidWorks Premium

A comprehensive 3D design solution adds to the capabilities of SolidWorks Professional with powerful simulation and design validation, as well as advanced wire and pipe routing functionality.

将一个综合的 3D 设计解决方案增加到 SolidWorks 专业内容中，这将使仿真、设计验证、线路及管道功能大大加强。

SolidWorks Professional

SolidWorks Professional builds on the capabilities of SolidWorks Standard to increase innovation and productivity, with data management, photo realistic rendering, and a sophisticated components and parts library.

为提高设计生产效率，SolidWorks Professional 植根于 SolidWorks Standard 的功能。该产品引入了数据管理、照片级逼真的效果图和复杂的零部件和零件库。

SolidWorks Standard

Unlock the benefits of this powerful 3D design solution for rapid creation of parts, assemblies, and 2D drawings with minimal training. Application-specific tools for sheet metal, weldments, surfacing, and mold tool and die make it easy to deliver best-in-class designs.

一款强大的 3D 设计解决方案，可让您通过最短的培训快速创建零件、装配体和 2D 工程图。要创建钣金、焊件、曲面、工模具和冲模，可使用应用程序特定的工具，这些工具可让您轻松创作出一流的设计。使用 SolidWorks Standard 可充分利用 3D 带来的好处。

SolidWorks offers a suite of simulation packages to set up virtual real-world environments so you can test your product designs before manufacture. Test against a broad range of parameters during the design process—like durability, static and dynamic response, motion of assembly, heat transfer, fluid dynamics, and plastics injection molding—to evaluate design performance and improve quality and safety. Simulation lowers costs and speeds time to market by reducing the number of physical prototypes you need before going into production. SolidWorks Simulation helps designers and engineers to innovate, testing and developing new concepts with greater insights.

SolidWorks 提供了一套仿真软件包，使用该软件包可以设置虚拟真实环境，以便您能够在制造之前测试产品设计。在整个设计过程中对很多参数（如持久性、静态和动态响应、装配体的运动、传热和流体力学）进行测试，以评估产品性能并做出决策来提高质量和安全。利用仿真可减少投产之前所需的物理样机数量，降低成本并加快上市速度。SolidWorks Simulation 可帮助设计师和工程师以更深入的洞察力创新、测试和开发新概念。

SolidWorks Simulation Premium

SolidWorks Simulation Premium software provides a full range of simulation capabilities to ensure product robustness and bolsters the depth of Simulation Professional with additional features, including tools for simulating nonlinear and dynamic response and dynamic loading.

SolidWorks Simulation Premium 软件提供了各种仿真功能来确保产品的可靠性，并通过附加功能增强 Simulation Professional 的深度，包括用于对非线性和动态响应以及动态载荷进行仿真的工具。

SolidWorks Simulation Professional

SolidWorks Simulation Professional goes beyond core simulation and expands the virtual test environment to product durability and natural frequencies, heat transfer and buckling, and pressure analysis and complex loading.

SolidWorks Simulation Professional 不仅包含核心仿真功能，还将虚拟测试环境扩展到产品耐用性和自然频率、传热、扭曲、压力分析和复杂载荷分析上。

SolidWorks Simulation

Included with the SolidWorks Premium 3D CAD design package, SolidWorks Simulation provides core simulation tools to test for strength and safety, analyze assembly kinematics, and simulate product performance to help you make the decisions that improve the quality.

SolidWorks Simulation 随附在 SolidWorks Premium 3D CAD 设计软件包中，提供核心仿真工具来测试强度和安全、分析装配体运动学和仿真产品性能，从而帮助您做出改进质量的决策。

SolidWorks Flow Simulation

SolidWorks Flow Simulation takes the complexity out of computational fluid dynamics to quickly and easily simulate fluid flow, heat transfer, and fluid forces critical to your design.

SolidWorks Flow Simulation 去除了计算流体力学分析的复杂性，可轻松快捷地仿真对设计至关重要的流体流动、传热和流体作用力。

SolidWorks SimulationXpress

SolidWorks SimulationXpress is a first-class basic stress analysis tool that comes with every SolidWorks 3D CAD software package.

SolidWorks SimulationXpress 是一款一流的基本应力分析工具，该工具随附在每个 SolidWorks 3D CAD 软件包中。

SolidWorks FloXpress

SolidWorks FloXpress is a first-class basic fluid flow analysis tool that calculates how water or air flows through part or assembly models and comes with every SolidWorks 3D CAD software package.

SolidWorks FloXpress 是一款一流的基本流体分析工具，可计算出水或气流通过零件或装配体模型的状态，该工具随附在每个 SolidWorks 3D CAD 软件包中。

SolidWorks Plastics

For companies that design plastic parts or injection molds, SolidWorks Plastics simulation software helps users predict and avoid manufacturing defects during the earliest stages of part and mold design, eliminating costly mold rework, improving part quality, and decreasing time to market.

对于进行塑料产品设计和注射模具设计的公司而言，Solidworks 塑料仿真软件能够帮助用户在零件和模具设计的早期阶段进行预测并避免制造缺陷，消除模具返工的成本，改善零件质量，并减少推向市场的时间。

(9) Introduction of SolidEdge

SolidEdge is a 3D CAD software owned by Siemens PLM, using Siemens PLM Software a companies have their own patented Parasolid software as a core, will popularize the CAD system and the world's leading entity modeling engines together, is based on the Windows platform, powerful and easy to use.

It supports the top of bottom of the design idea and its functions such as core of modeling, sheet metal design, assembly design, products manufacturing information management, production plans, value chains, embedded finite element and product data management. These functions take leading posinion that those of the similar software. The core of design is the best choice for personnel and business and has been successfully applied to machinery electronics, aerospace, automotive, instrumentation, mold, shipbuilding and other consumer goods industry. At the same time the system also provides the two-dimensional to three-dimensional solid conversion tools. You need not abandon many-years drawing results when Solid Edge quickly jumps to 3D design. This qualitative leap allows you to experience the superiority of 3D design.

SolidEdge 简介

SolidEdge 是三维 CAD 软件，归 Siemens PLM 所有，使用该公司自己拥有专利的 ParaSolid 作为软件核心。SolidEdge 的使用将使 CAD 系统与处世界领先地位的实体模型机同样受欢迎。该软件基于 Window 平台、功能强大、操作简便。

它支持自上向下和自下向上的设计理念，涵盖范围极广，包括模型核心、钣金设计、装配设计、产品制造、信息管理、生产计划、价值链、嵌入式有限模块分析以及生产数据管理，与同类软件相比，功能处于领先地位。设计的核心理念是个人、企业发展的首选，并且已成功应用于电子、航空、汽车、仪器、模具、造船、消费品等行业。同时系统还提供了从二维视图到三维实体的转换工具，您无需摒弃多年来二维制图的成果，借助 Solid Edge 就能迅速跃升到三维设计，这种质的飞跃能让您体验到三维设计的巨大优越性。

(10) Introduction of GstarCAD

HaoChen CAD is a CAD software product developed and researched independently by Suzhou Hao-chen Software Co., Ltd. Flagship product Chen CAD platform, known as the "design Office software", is widely used in construction, manufacturing and other fields of design. Chen CAD platform software has, including Chinese, traditional Chinese, English, Japanese, Russian, Korean, Germany French, Spanish, Hebrew, 13 language version.

GstarCAD 简介

浩辰 CAD 是苏州浩辰软件股份有限公司自主研发的 CAD 软件产品。旗舰产品浩辰 CAD 平台，被喻为"设计领域的 Office 软件"，广泛应用于工程建筑、制造业等设计领域。浩辰 CAD 平台软件目前已拥有简体中文、繁体中文、英文、日文、俄文、韩文、德文、法文、西班牙语、希伯来语等 13 个语言版本。

(11) Introduction of UG NX

NX is a next-generation digital product development system that helps companies transform the product lifecycle. With the industry's broadest suite of integrated, fully associative CAD/CAM/CAE applications, NX touches the full range of development processes in product design, manufacturing and simulation.

NX provides a complete suite of integrated process automation tools to enable companies to capture and reuse product and process knowledge encouraging the use of corporate best practice.

UG NX 简介

NX 是下一代的数字化产品开发系统，旨在帮助用户改变产品生命周期。它集成工业级的最广泛的套件，实现 CAD/CAM/CAE 应用全相关，NX 触及产品设计、制造和仿真中全方位的开发过程。

NX 提供了一个完整的工业自动化集成套件，能够帮助公司捕获和重新运用产品和流程知识，鼓励公司实现应用最佳化。

四、常用软件介绍 127

Industrial Design & Styling

　　NX delivers a complete set of flexible shape creation, manipulation, and analysis tools and is an integrated part of a complete digital product development solution.

　　NX 提供了一套完整的形状创造、操纵和分析工具,它是整个数字化产品开发方案中的一个组成部分。

Package Design

　　NX software is an ideal tool for the challenges faced by package designers in the consumer packaged goods and food & beverage industries.

　　NX 软件是一个理想的工具,它完全胜任消费性包装商品和餐饮业所面临的挑战。

Mechanical Design

　　NX mechanical design tools deliver superior power, productivity, flexibility, and coordination for product development.

　　NX 机械设计工具为产品开发提供卓越的动力、生产力、灵活性和协调性。

Electromechanical Systems Design

　　NX streamlines and accelerates electromechanical systems design with a solution that integrates mechanical, electrical, and electronic components.

　　NX 简化并加速了机电系统的设计,它集成了机械、电器和电子元器件等。

Visual Reporting & Analytics

NX delivers visual reporting and analytics with High Definition 3D (HD3D) technology to instantly gather PLM data and visualize its impact directly in the context of the 3D design.

NX 提供了高清 3D(HD3D) 技术的可视化报告和分析工具,它即时收集 PLM 数据,并进行可视化,直接影响了 3D 设计环境。

Mechanical Simulation

NX provides the industry's broadest range of multi-discipline simulation solutions that leverage powerful capabilities in model preparation, solving, and post processing.

NX 为业界提供了范围最广的多学科仿真技术方案,它在模型预处理、求解以及后期处理方面具有强大的能力。

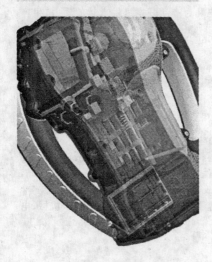

Electromechanical Simulation

NX includes simulation solutions for all of the primary failure modes for electromechanical products: temperature, vibration, and dust or humidity.

NX 涵盖了所有机电产品主要故障模式的仿真方案:温度、振动以及灰尘或湿度。

Tooling & Fixture Design

NX Tooling applications extend design productivity and efficiency into manufacturing, dynamically linking to product models for accurate tooling, molds, dies, and workholding fixtures.

NX 工装设计应用环境把设计生产能力和效率延伸到制造领域,把精确的刀具、塑料模具、冲压模具以及工件装夹夹具动态地连接到产品上。

Machining

NX CAM delivers a complete solution for machine tool programming that maximizes throughout of the most advanced machine tools.

NX CAM 为机床编程提供了完整的解决方案,它能够使最先进的机床发挥出最大的效益。

Engineering Process Management

NX engineering process management, powered by Teamcenter, provides a single source of product engineering and process knowledge, seamlessly integrated with CAD, CAM, and CAE.

NX 工程过程管理由 Teamcenter 支持,提供单一的产品工程、工艺知识源,与 CAD、CAM 和 CAE 无缝集成。

Inspection Programming

NX CMM Inspection Programming enables you to automate programming to save time and improve accuracy.

NX CMM 测量编程工具能够进行自动化的程序设计以利于节省时间,并提高精度。

Mechatronics Concept Designer

Mechatronics Concept Designer helps you design machines with greater speed and quality by providing a complete, end-to-end solution that enables easier collaboration between disciplines, reuse of existing knowledge, and better decision making through concept evaluation.

机电一体化概念设计器帮助你以更快的速度、更高的质量来设计机器,通过概念评估,提供一个完整的、端对端的解决方法,使多学科知识之间更易于协同、已有的知识更易于重用,并且能够做出更好的市场决策。

(12) Introduction of Creo Elements/Pro

A product formerly known as Pro/ENGINEER is a parametric, integrated 3D CAD/CAM/CAE solution created by Parametric Technology Corporation (PTC). It was the first to market with parametric, feature-based, associative solid modeling software. The application runs on Microsoft Windows platform, and provides solid modeling, assembly modelling and drafting, finite element analysis, and NC and tooling functionality for mechanical engineers. The Pro/ENGINEER name was changed to Creo Elements/Pro on October 28, 2010, coinciding with PTC's announcement of Creo, a new design software application suite.

Creo Elements/Pro 简介

其前身是著名的 Pro/ENGINEER,它是由参数化技术公司(PTC)开发的一款集成的三维 CAD/CAM/CAE 解决方案。它是第一款推向市场的集参数化、基于特征、相关联的实体于一体的建模软件。它运行于 Windows 平台,为机械工程师提供实体建模、装配建模、工程图、有限元分析、NC 以及工装等功能。2010 年 10 月 28 日 Pro/ENGINEER 更名为 Creo Elements/Pro,同时,PTC 声明 Creo 为一套新的设计软件。

Overview 概述

Creo Elements/Pro (formerly Pro/ENGINEER), PTC's parametric, integrated 3D CAD/CAM/CAE solution, is used by discrete manufacturers for mechanical engineering, design and manufacturing.

Created by Dr. Samuel P. Geisberg in the mid-1980s, Pro/ENGINEER was the industry's first successful rule-based constraint (sometimes called "parametric" or "variational") 3D CAD modeling system. The parametric modeling approach uses parameters, dimensions, features, and relationships to capture intended product behavior and create a recipe which enables design automation and the optimization of design and product development processes. This design approach is used by companies whose product strategy is family-based or platform-driven, where a prescriptive design strategy is fundamental to the success of the design process by embedding engineering constraints and relationships to quickly optimize the design, or where the resulting geometry may be complex or based upon equations. Creo Elements/Pro provides a complete set of design, analysis and manufacturing capabilities on one integral, scalable platform. These required capabilities include Solid Modeling, Surfacing, Rendering, Data Interoperability, Routed Systems Design, Simulation, Tolerance Analysis, and NC and Tooling Design.

Companies use Creo Elements/Pro to create a complete 3D digital model of their products. The models consist of 2D and 3D solid model data which can also be used downstream in finite element analysis, rapid prototyping, tooling design, and CNC manufacturing. All data is associative and interchangeable between the CAD, CAE and CAM modules without conversion. A product and its entire bill of materials (BOM) can be modeled accurately with fully associative engineering drawings, and revision control information. The associativity functionality in Creo Elements/Pro enables users to make changes in the design at any time during the product development process and automatically update downstream deliverables. This capability enables concurrent engineering — design, analysis and manufacturing engineers working in parallel — and streamlines product development processes.

Creo Elements/Pro is part of a broader product development system developed by PTC. It connects to PTC's other solutions that aid product development, including Windchill, ProductView, MathCAD and Arbortext.

Creo Elements/Pro 的(前身为 Pro/ENGINEER)，PTC 的参数化、集成式 3D CAD/CAM/CAE 技术解决方案，用于离散型制造企业的机械工程、设计和制造方面。

在 20 世纪 80 年代中期由 Samuel P. Geisberg 创建的 Pro/ENGINEER 是业界第一个成功使用基于规则的约束(有时称为"参数化"或"变量")的三维 CAD 建模系统。参数化建模方法使用参数、尺寸、特征以及关系来捕捉预期的产品行为，并创建能够使设计自动化、设计优化以及推进产品开发流程的方法。公司使用这种设计方法，它的产品战略是基于家庭或平台驱动的，规范的设计策略是设计过程成功的基础，它通过嵌入工程约束和关系以迅速优化设计，完成复杂造型或基于方程的造型。Creo Elements/Pro 在一个集成的可扩展的平台上提供了一整套设计、分析和制造功能。这些必需的功能包括实体建模、曲面建模、渲染、数据互操作、布线系统、仿真、公差分析以及 NC 和工装设计。

公司使用 Creo Elements/Pro 创建一个完整的三维数字模型。他们的产品包括二维和三维实体模型的数据，也可用于下游的有限元分析、快速成型、工装设计和 CNC 制造。所有数据无需转换，在 CAD、CAE 和 CAM 之间实现关联和互换。产品和它的全部材料清单(BOM)使用完全关联的工程图纸以及版本控制信息，可以实现精确的建模。Creo Elements/Pro 的关联功能能够使用户在产品设计过程的任何时间内更改设计，并自动更新下游产品。这种能力使并行工程变得可行，设计、分析和制造工程师并行工作并且简化产品开发流程。

Creo Elements/Pro 是 PTC 开发得更广泛的产品开发系统的一部分，它连接到 PTC 的其他解决方案辅助产品开发，这些产品包括 Windchill、ProductView、MathCAD 和 Arbortext。

Summary of capabilities　能力概要

Like any software it is continually being developed to include new functionality. The details below aim to outline the scope of capabilities to give an overview rather than giving specific details on the individual functionality of the product.

Creo Elements/Pro is a software application within the CAID/CAD/CAM/CAE category, along with other similar products currently on the market.

Creo Elements/Pro is a parametric, feature-based modeling architecture incorporated into a single database philosophy with advanced rule-based design capabilities. It provides in-depth control of complex geometry, as exemplified by the trajpar parameter. The capabilities of the product can be split into the three main headings of Engineering Design, Analysis and Manufacturing. This data is then documented in a standard 2D production drawing or the 3D drawing standard ASME Y14.41-2003.

像其他软件不断发展一样，该软件包含着新的功能。下面的详细信息旨在勾勒出能力范围的一个概要，而不是给出个别产品功能的具体细节。

Creo Elements/Pro 和目前市场上其他类似的产品一样，是一款 CAID/CAD/CAM/CAE 类的应用软件。

Creo Elements/Pro 是一款参数化、基于特征的建模体系，引入单一数据库理念，具有高级的基于规则的设计能力。它作为 trajpar 参数化的典范，提供复杂几何体的深入控制。产品的能力可以分为三个主题：设计、分析和制造。然后，这些数据归档为标准的 2D 产品工程图或 3D 绘图标准 ASME Y14.41-2003。

Engineering Design　**工程设计**

Creo Elements/Pro offers a range of tools to enable the generation of a complete digital representation of the product being designed. In addition to the general geometry tools there is also the ability to generate geometry of other integrated design disciplines such as industrial and standard pipe work and complete wiring definitions. Tools are also available to support collaborative development.

A number of concept design tools that provide up-front Industrial Design concepts can then be used in the downstream process of engineering the product. These range from conceptual Industrial design sketches, reverse engineering with point cloud data and comprehensive free-form surface tools.

Creo Elements/Pro 提供一系列的工具,能够使设计的产品进行完整的数字表现。除了一般几何工具外,它也能产生其他综合学科的几何体,比如,工业级标准管道作业和完整的布线定义等。它也可以作为系统开发的支撑。

许多提供前期工业设计理念的概念设计工具可以用在工程产品的下游过程。这些范围涵盖工业设计概念草图、逆向工程点云数据和全面的自由曲面工具。

Analysis　**分析**

Creo Elements/Pro has numerous analysis tools available and covers thermal, static, dynamic and fatigue finite element analysis along with other tools all designed to help with the development of the product. These tools include human factors, manufacturing tolerance, mould flow and design optimization. The design optimization can be used at a geometry level to obtain the optimum design dimensions and in conjunction with the finite element analysis.

Creo Elements/Pro 拥有众多的分析工具,并涵盖热、静态、动态和疲劳有限元分析,和其他工具一起帮助产品的开发。这些工具包括人为因素、制造公差、模内流动和优化设计。优化设计可用于几何水平,以获得最佳的设计尺寸,并结合有限元分析进行优化设计。

Manufacturing　**制造**

By using the fundamental abilities of the software with regards to the single data source principle, it provides a rich set of tools in the manufacturing environment in the form of tooling design and simulated CNC machining and output.

Tooling options cover specialty tools for molding, die-casting and progressive tooling design.

通过使用软件的基本能力以及访问单一数据源的原理,它在制造环境,诸如工装设计、CNC 加工仿真和输出等方面提供了一套丰富的工具。

工装选项涵盖用于注塑、铸造以及级进模具设计工具。

(13) Introduction of CATIA

CATIA is a high-end CAD/CAM software developed by France's Dassault aircraft company. CATIA has a high reputation with its powerful surface design features in the design of airplanes, cars, ships. CATIA surface modeling capabilities provides a very rich modeling tool to support the modeling needs of the user. Such as its unique high surface functionality of Bezier curves, the number can reach 15. The demanding requirements of surface smoothness in special industry can be met.

CATIA is the abbreviation of the English Computer Aided Tri-Dimensional Interface Application. It is a mainstream CAD/CAE/CAM integrated software in the world. CATIA is the flagship product of the French Dassault company's product development solution. As an important component of PLM collaborative solutions, it can help manufacturers design their products, and support from the pre-project phase, the specific design, analysis, simulation, assembly and maintenance, including all the industrial design process.

The modular CATIA series of products are designed to meet the needs of customers in the product development activities, including the style and shape design, mechanical design, equipment and systems engineering, managing digital prototyping, machining, analysis and simulation. CATIA products are based on open and scalable V5 architecture.

By enabling enterprises to reuse product design knowledge, shorten development cycles, CATIA solutions to accelerate business response to market demand. Since 1999, the market is widely used in the digital prototyping process, thus making it the world's most commonly used in product development system.

The CATIA series in seven areas have been to become the first 3D design and simulation solutions: automotive, aerospace, shipbuilding, plant design, power and electronics, consumer goods and general machinery manufacturing.

CATIA 简介

CATIA 是法国达索飞机公司开发的高档 CAD/CAM 软件。CATIA 软件以其强大的曲面设计功能而在飞机、汽车、轮船等设计领域享有很高的声誉。CATIA 的曲面造型功能提供了极丰富的造型工具来支持用户的造型需求。比如其特有的高次 Bezier 曲线曲面功能，次数能达到 15，能满足特殊行业对曲面光滑性的苛刻要求。

CATIA 是英文 Computer Aided Tri-Dimensional Interface Application 的缩写，是一种主流的 CAD/CAE/CAM 一体化软件。CATIA 是法国达索公司的产品开发旗舰解决方案。作为 PLM 协同解决方案的一个重要组成部分，它可以帮助制造厂商设计他们未来的产品，并支持从项目前阶段、具体的设计、分析、模拟、组装到维护在内的全部工业设计流程。

模块化的 CATIA 系列产品旨在满足客户在产品开发活动中的需要，包括风格和外形设计、机械设计、设备与系统工程、管理数字样机、机械加工、分析和模拟。CATIA 产品基于开放式可扩展的 V5 架构。

通过使企业能够重用产品设计知识、缩短开发周期，CATIA 解决方案加快企业对市场需求的反应。自 1999 年以来，市场上广泛采用它的数字样机流程，从而使之成为世界上最常用的产品开发系统。

CATIA 系列产品已经在七大领域里成为首要的 3D 设计和模拟解决方案：汽车、航空航天、船舶制造、厂房设计、电力与电子、消费品和通用机械制造。

CATIA-Advanced hybrid modeling techniques　CATIA 先进的混合建模技术

Hybrid modeling of design objects: In CATIA design environment, whether solid or surface, it achieves true interoperability;

设计对象的混合建模：在 CATIA 的设计环境中，无论是实体还是曲面，都做到了真正的互操作；

Variables and parameters of the hybrid modeling: At design time, designers do not have to consider how to parameterize design goals, CATIA provides a variable driven and post parameterized capabilities.

变量和参数化混合建模：在设计时，设计者不必考虑如何参数化设计目标，CATIA 提供了变量驱动及后参数化能力。

Hybrid modeling of geometry and intelligent works: For a business enterprise, many years of experience accumulated to a knowledge base of CATIA is used to guide the novice of the enterprise, or to accelerate new models to market as soon as possible.

几何和智能工程混合建模：对于一个企业，可以将企业多年的经验积累到 CATIA 的知识库中，用于指导本企业新手，或指导新型号的开发，以加速新型号推向市场的时间。

CATIA facilitates to change within the entire product life cycle, especially in the late modification. CATIA 具有在整个产品周期内的便捷的修改能力，尤其是在后期修改。

Either solid modeling or surface modeling, CATIA provides intelligent tree structure. Users can be convenient and efficient to repeat product modifies, even major changes in the final stages of the design needed to be done, and the original program upgraded are very easy things for CATIA.

无论是实体建模还是曲面造型，由于 CATIA 提供了智能化的树结构，用户可方便快捷地对产品进行重复修改，即使是在设计的最后阶段需要做重大的修改，或者是对原有方案的更新换代，对于 CATIA 来说，都是非常容易的事。

All modules of CATIA with full Associative 所有具有全相关性 CATIA 的模块

As CATIA modules based on a unified data platform, there is true correlation in each module of CATIA. Three-dimensional model can be fully reflected in the two-dimensional and finite element analysis, mold and CNC machining process.

CATIA 的各个模块都基于统一的数据平台,因此 CATIA 的各个模块均存在着真正的全相关性。三维模型的修改能完全体现在二维分析以及有限元分析、模具和数控加工的过程中。

Concurrent engineering design environment allows designers to shorten the cycle
并行工程的设计环境使得设计周期大大缩短

CATIA working environment of the multi-model links and hybrid modeling approach, making concurrent engineering design model is no longer a new concept, as long as the overall design department distributes out the basic structure size, each subsystem can start to work: work together, not interrelated. With the interconnection between models, the upstream design results can be used as the reference of the downstream, and design changes from upstream can directly affect the refresh of the downstream work: to achieve a true concurrent engineering design environment.

CATIA 提供的多模型链接的工作环境及混合建模方式,使得并行工程设计模式已不再是新鲜的概念,总体设计部门只要将基本的结构尺寸发放出去,各分系统的人员便可开始工作,既可协同工作,又不互相牵连;由于模型之间的互相联结性,使得上游设计结果可作为下游的参考,同时,上游对设计的修改能直接影响到下游工作的刷新,从而实现真正的并行工程设计环境。

CATIA covering the entire process of the product development CATIA 覆盖了产品开发的整个过程

CATIA provides a complete design capability: from product concept design to final product formation, with its accurate and reliable solution to provide a complete 2D and 3D parametric hybrid modeling and data management tools, from the design of individual parts to the final electronic prototype; at the same time, as a fully integrated software system, CATIA has combined mechanical design, engineering analysis and simulation, CNC machining with CAT web network application solutions organically to provide users with strictly paperless work environment, especially industry-specific modules for cars, motorcycles in CATIA. CATIA has a broad professional coverage to help customers shorten the design and production cycles, improve product quality and reduce the costs.

CATIA 提供了完备的设计能力:从产品的概念设计到最终产品的形成,以其精确可靠的解决方案提供了完整的 2D、3D 参数化混合建模及数据管理手段,从单个零件的设计到最终电子样机的建立;同时,作为一个完全集成化的软件系统,CATIA 将机械设计、工程分析及仿真、数控加工和 CAT 网络应用解决方案有机地结合在一起,为用户提供严密的无纸工作环境,特别是 CATIA 中的针对汽车、摩托车业的专用模块,使 CATIA 拥有了宽广的专业覆盖面,从而帮助客户达到缩短设计生产周期、提高产品质量及降低费用的目的。

(14) Introduction of CAXA

CAXA ten years to adhere to the "software serving manufacturing" concept, developed over 20 series of software products, covering the informational design of manufacturing industry, technology, manufacturing and management of four major areas, has won "Native Ten Top Software" for five consecutive years and the honor of China Software Industry Association the "Golden Software Award" for 20 years; CAXA always insists on market-oriented path, and it has accumulated sales of genuine software more than 150,000 sets by the end of 2004, and won the trust and praise of the majority of business users and engineering and technical personnel; also successfully established a technical service system consisting of 35 offices throughout the country, more than 300 education and training centers, more than 300 agents and dealers and multi-level partners, as the CAD/CAM/PLM industry leader and major supplier in China.

The four letters of CAXA: C (computer) - A (aided), X (any) - A (alliance, ahead), "one step ahead of computer-aided technologies and services" (Computer Aided X Alliance - Always a step Ahead).

CAXA 简介

CAXA 十多年来坚持"软件服务制造业"理念,开发出二十多个系列的软件产品,覆盖了制造业信息化设计、工艺、制造和管理四大领域,曾连续五年荣获"国产十佳优秀软件"以及中国软件行业协会 20 年"金软件奖"等荣誉;CAXA 始终坚持走市场化的道路,截至 2004 年底已累计成功销售正版软件超过 150 000 套,赢得广大企业用户与工程技术人员的信任和好评;还成功在全国建立起了 35 个办事处、300 多个教育培训中心、300 多家代理经销商和多层次合作伙伴组成的技术服务体系,是我国 CAD/CAM/PLM 业界的领导者和主要供应商。

CAXA 是由:C——Computer(计算机),A——Aided(辅助的),X(任意的),A——Alliance、Ahead(联盟、领先)这四个字母组成的,其涵义是"始终领先一步的计算机辅助技术和服务"(Computer Aided X Alliance - Always a Step Ahead)。

(15) Introduction of Inventor

Inventor of U.S. Autodesk Company has launched a three-dimensional visualization of physical simulation software Autodesk Inventor Professional (AIP), has now introduced the latest version of AIP2013, also introduced the iphone version, on sale in the app store. The Autodesk Inventor Professional, Autodesk Inventor & 3D design software; developed AutoCAD-based platform for 2D mechanical drawing and detailing software AutoCAD Mechanical; also joined for cable and wire harness design, pipeline design and PCB IDF file input function module, and joined the technical support of industry-leading ANSYS FEA that stress analysis can be directly in Autodesk Inventor. On this basis, the integrated data management software Autodesk Vault- is used to safely manage the design data in progress. Autodesk Inventor Professional set of all these products in one, thus providing a risk-free two-dimensional to three-dimensional conversion path. Now, you can convert to three-dimensional by your own pace, the protection of two-dimensional graphics and investment in knowledge, and clearly knew to use the plat form with the strongest compatibility in the current market.

Autodesk Inventor software is a comprehensive set of design tools used to create and validate a complete digital prototype; help manufacturers to reduce the cost of physical prototyping and bring more innovative products to market morequickly.

Autodesk Inventor product line is changing the traditional CAD workflow: to simplify the creation of complex 3D models, engineers can focus on the functionality of the design. And use the digital prototype to quickly create a digital prototype to validate the functionality of the design, engineers can be easier to detect design errors before production. Inventors can speed up the concept design to manufacturing process, and by virtue of this innovative approach, for seven consecutive years sales rank first among similar products.

Inventor 简介

Inventor 美国 AutoDesk 公司推出的一款三维可视化实体模拟软件 Autodesk Inventor Professional(AIP),目前已推出最新版本 AIP2013。同时还推出了 iphone 版本,在 app store 有售。Autodesk Inventor Professional 包括 Autodesk Inventor& 三维设计软件;基于 AutoCAD 平台开发的二维机械制图和详图软件 AutoCAD Mechanical;还加入了用于缆线和束线设计、管道设计及 PCB IDF 文件输入的专业功能模块,并加入了由业界领先的 ANSYS 技术支持的 FEA 功能,可以直接在 Autodesk Inventor 软件中进行应力分析。在此基础上,集成的数据管理软件 Autodesk Vault- 用于安全地管理进展中的设计数据。由于 Autodesk Inventor Professional 集所有这些产品于一体,因此提供了一个无风险的二维到三维转换路径。现在,您能以自己的进度转换到三维,保护现在的二维图形和知识投资,并且清楚地知道自己在使用目前市场上 DWG 兼容性最强的平台。

  Autodesk Inventor 软件是一套全面的设计工具,用于创建和验证完整的数字样机;帮助制造商减少物理样机投入,以更快的速度将更多的创新产品推向市场。

  Autodesk Inventor 产品系列正在改变传统的 CAD 工作流程:因为简化了复杂三维模型的创建,工程师即可专注于设计的功能实现。通过快速创建数字样机,并利用数字样机来验证设计的功能,工程师可在投产前更容易发现设计中的错误。Inventor 能够加速从概念设计到产品制造的整个流程,并凭借着这一创新方法,连续 7 年销量居同类产品之首。

(16) Introduction of AutoCAD

  AutoCAD (Auto Computer Aided Design) is the first production of automatic computer aided design software of the United States of America Autodesk company in 1982, used for 2D drawing; detailed drawing, design documents and basic 3D design. Now it has become an internationally popular drawing tool. AutoCAD has a good user interface, through the interactive menus or the command lines it can carry out various operations. As its design environment of multiple documents, non computer professionals can quickly learn to use. In the process of continuous practice users can grasp its various applications and the skills betwer of development, so as to continuously improve work efficiency. AutoCAD has a wide adaptability; it can be used in a variety of operating system support microcomputers and workstations.

AutoCAD 简介

  AutoCAD(Auto Computer Aided Design) 是美国 Autodesk 公司首次于 1982 年生产的自动计算机辅助设计软件,用于二维绘图、详细绘制、设计文档和基本三维设计。现已经成为国际上广为流行的绘图工具。AutoCAD 具有良好的用户界面,通过交互菜单或命令行方式便可以进行各种操作。它的多文档设计环境让非计算机专业人员也能很快地学会使用。在不断实践的过程中更好地掌握它的各种应用和开发技巧,从而不断提高工作效率。AutoCAD 具有广泛的适应性,它可以在各种操作系统支持的微型计算机和工作站上运行。

五、面试情景对话
5. Situational Dialognes at Interviews

—Susan: May I have your name?
—Susan: 您叫什么名字?
—Tom: My name is Tom.
—Tom: 我叫汤姆。
—Susan: Nice to meet you, Tom. I'm Susan, HRD (Human Resource Director) of the company. Please have a seat.
—Susan: 很高兴认识你,Tom。我是人力资源总监 Susan,请坐。
—Tom: Nice to meet you, Ms. Susan.
—Tom: 您好,Susan 女士。
—Susan: Tell me something about yourself, please.
—Susan: 您能介绍一下自己的情况吗?
—Tom: My major is Mechanical Engineering. During my 4 years in the university, I have passed all the professional courses with excellent results such as Principles of Electric Circuits, Machine Principles, Machine Design, Mechanics of Materials, Theoretical Mechanics and so on. Although I'm not the best in my class according to the grades of all the subjects, I have my advantages. I won the first prize at the Mechanical Design Contest. I also got a grade A in Major Course Design. I got an excellent in Graduation Thesis. So you can see I am the best in terms of experimental and practical abilities. I am also good at programming both assembly language and advanced languages such as C++ and VB. So I think I'm suitable for this job.
—Tom: 我的专业是机械工程。在大学的四年中,我以优异成绩通过所有的专业课,如电路原理,机械原理,机械设计,材料力学,理论力学等。尽管按全部成绩来看我在班级里并不是最好的,但我有自己的优势。我曾赢得机械设计大赛一等奖,在专业课程设计中获得优秀,毕业设计论文也是优秀。由此您可以看出在实验能力和动手能力上我是最好的。我还善于用汇编语言和高级语言如 C++ 和 VB 编程。因此我觉得我很适合这项工作。
—Susan: Good. Do you have any hobbies?
—Susan: 好的,你有什么爱好吗?
—Tom: Of course. I'm good at table tennis, and I'm the point guard of the Department Basketball Team. We once won the championship in the university basketball contest.
—Tom: 当然,我乒乓球打得不错,还是系篮球队的控球后卫,我们获得过学院冠军。
—Susan: Really?
—Susan: 真的吗?
—Tom: Yes, I also like singing and traveling.
—Tom: 是的,我还喜欢唱歌和旅行。
—Susan: That's impressive. From your talk, I can see that you have good performance in your university. What I want to know is why you want to choose this job.
—Susan: 很不错。从你的谈话,我可以看出你在校时很优秀。我想知道为什么你想到这里来工作。
—Tom: I have paid close attention to the news from your industry, and deeply impressed by the development of your company. As I know, TMD Company is a joint venture and is the top 5 in this field in China. In the past several years, many excellent undergraduate students have been employed in your company. I talked with some of them, so I know something about your company. I also like your

enterprise culture, for example, the employees' training and development plan. I find that my personal goals and ideas about business operations fit perfectly with the company's. I am confident that I will have a very rewarding and successful career here.

— Tom：我一直关注你们行业的新闻，贵公司的发展给我留下了很深的印象。据我所知 TMD 公司是一个合资企业，在国内位于该领域前五。在过去几年中，有很多优秀的大学毕业生来到贵公司。我和他们中的一些人交谈过，对贵公司有一些了解。同时我也很喜欢贵公司的企业文化如员工的培训及发展计划。我觉得我的个人发展目标、商业理念和公司的目标完全吻合。我相信我在这儿能开创有价值而且成功的事业。

— Susan: The sense of responsibility to the company and the team is very important for us. Do you have any ideas about that?

— Susan：在我们这里，对公司以及团队的责任是非常重要的。关于这一点，你有什么想法？

— Tom: I am very responsible. My teacher and classmates would always count on me to complete a certain task. I am sure what I ought to do. I get along well with people and enjoy working with others.

Tom：我很有责任心，如果让我负责一件事，老师和同学们都会很放心。我确定我该做什么。我擅于与人相处，喜欢和他人一起工作。

— Susan: So what are your career objectives? And what do you think of your own development if you get the job?

— Susan：那你的事业目标是什么呢？如果你获得了这个职位，你对自己的发展有什么想法？

— Tom: I hope to have a good opportunity to put all of my knowledge into practice. I'd like to start with the research and development work. I am a doer and believe I can contribute a great deal to the company. I hope that in a couple of years, I could manage my work independently and be able to lead an energetic and efficient team.

— Tom：我希望得到很好的实战机会。我想从研发工作做起，我是个实干者，我相信自己能为公司贡献很多。我希望在几年内能够独立承担自己的工作并且领导一支有活力并高效的团队。

— Susan: Good, then do you have any questions about our company?

— Susan：好的，对于公司，你有什么问题吗？

— Tom: Could you please tell me something about your training program?

— Tom：能否介绍一下你们的培训制度？

— Susan: In brief, we offer both in—house and off—site training to U.S. head quarters. We have a few daylong training sessions for topics like business writing skills and software training etc.. These sessions are available to everyone who applies. We also have a variety of other programs based on each work function. Our program is mainly a job—rotation program, and we believe it is more effective than traditional on—the—job training.

— Susan：简单来说，我们既有在岗培训，也有去美国总部专门培训。例如，我们采用整日课程来进行商务写作和软件操作一类的培训，这样的培训每个人都可以申请参加。我们还针对各个职能安排了多种多样的课程。我们的课程主要是轮岗，我们相信这比传统的在职培训要有效得多。

— Tom: It sounds quite attractive!

— Tom：听上去真的很吸引人啊！

— Susan: Lastly, what's your main consideration in your current job hunt?

— Susan：最后，目前你选择工作时主要考虑哪个方面？

— Tom: Potential growth opportunity. If a potential job does offer me an opportunity to grow and become mature, I will find it very easy to throw myself into my work and finish my responsibilities at the highest level. I have reasons to believe that your company will be a perfect place for me to grow and develop!

— Tom：潜在的发展机会。如果一个工作能够让我有发展和成熟的机会，我就会更容易地全身心投入去完成我的职责。我有理由相信贵公司将是我成长和发展的良好平台!
——Susan: I also hope to have the chance to get to know you better in the future.
——Susan: 好的,我也希望有机会更加详细地了解你。
——Tom: Thank you for your time. I am looking forward to working here.
——Tom: 谢谢您的时间。我期待着有机会来这里工作。
——Susan: You'll hear from us next week.
——Susan: 我们下周会通知你。
——Tom: Thank you very much. See you then.
——Tom: 非常感谢您!再见。

参考文献
References

[1] 朱冬梅,等.画法几何及机械制图.北京:高等教育出版社,2007
[2] 何铭新,钱可强.机械制图.北京:高等教育出版社,1998
[3] 王巍,等.机械工程图学.北京:机械工业出版社,2006
[4] 蒋寿伟,等.现代机械工程图学.北京:高等教育出版社,2006
[5] 魏莉,等.画法几何及机械制图补充习题集.校内印刷
[6] 廖念钊,等.互换性与技术测量.4版.北京:中国计量出版社,2000
[7] 甘永立.几何量公差与检测.4版.上海:上海科学技术出版社,1997
[8] 甘永立.几何量公差与检测习题试题集.4版.上海:上海科学技术出版社,2000
[9] 刘德全.语码转换式双语教学系列教材机电一体化技术.大连:大连理工大学出版社,2008
[10] 徐国凯.语码转换式双语教学系列教材电子与自动化技术.大连:大连理工大学出版社,2008
[11] 刘天模,徐幸梓.工程材料.北京:机械工业出版社,2001
[12] 王焕庭,等.机械工程材料.4版.大连:大连理工大学出版社,2003
[13] 史美堂.金属材料及热处理.上海:上海科学技术出版社,2000
[14] 机械工程材料——辅导、习题、实验.大连:大连理工大学出版社,2003
[15] 邓文英.金属工艺学.4版.北京:高等教育出版社,2000
[16] 金属工艺学.杭州:浙江大学出版社,2000

索 引
Index

(电阻)点焊 15-4
端口 7-3
[奥氏体]形变热处理 14-6
16 位机 7-2
4 位机 7-2
8 位机 7-2
s 平面 17-3
α-铁素体 14-4

A
阿基米德蜗杆 2-6
安培 12-4
安培表 18-3
安全离合器 3-5
安全销 3-2
安装 6-6
安装尺寸 1-9
奥匹兹编码系统 11-7
奥氏体 14-4
奥氏体不锈钢 14-7
奥氏体分解 14-6
奥氏体化 14-6
奥氏体晶粒度 14-6
奥氏体转变 14-6
奥氏体转变曲线 14-6

B
八叉树 11-5
八进制 7-1
八位机 7-5
巴比合金/巴氏合金 14-9
拔长 15-3
拔模 9-6
白口铸铁 14-8
百分表 10-2
百科全书 17-1
摆动从动件 2-5
摆动载荷 10-6
摆线齿轮 2-6
斑马条纹 9-8

板弹簧 3-5
板料成形 15-3
板牙 6-2
半波整流 18-4
半导体 18-2
半导体晶片 12-5
半对数坐标 17-5
半径标注 9-5
半剖视图 1-6
半圆键 1-7
半自动机床- 6-3
伴随矩阵 17-6
包角 3-3
包容要求 10-3
保持架 3-5
保留符号 7-5
保温时间 14-6
爆炸焊 15-4
爆炸视图 9-6
杯形砂轮 6-2
贝塞尔曲线 11-5
贝氏体 14-6
贝氏体转变 14-6
背吃刀量(切削深度) 6-2
被动测量 10-2
被动的,无源的 17-4
被动力 5-2
本构关系 4-2
崩碎切屑 6-2
比较法 10-4
比较仪 10-2
比较转移 7-4
比例 11-4
比例积分控制 17-4
比例极限 4-2
比例缩放 9-3
比例微分控制 17-4
闭环控制 17-1
闭环轮廓 9-8
闭式链 2-1
边界表示法 11-5

边界框 9-6
边界摩擦 3-1
边界约束 11-6
边沿触发 7-6
编辑尺寸 9-5
编辑属性 9-4
编码 11-7
编码系统 7-1
变刚度弹簧 3-5
变换 17-2
变截面梁 4-5
变力的功 5-13
变量 7-5
变量表 7-5
变容二极管 18-6
变矢量 5-11
变速箱 6-3
变位齿轮 2-6
变形 14-5
变形固体 4-1
变形抗力 15-3
变形现象 8-4
变形协调方程 4-2
变形织构 14-5
变压器 18-2
遍历 11-3
遍历树 11-3
标尺 10-2
标号 7-5
标量 17-6
标题栏 1-1
标注尺寸的基本要求 1-4
标注和查表 1-8
标注特征比例 9-5
标注形位公差 9-5
标注样式 9-5
标注样式管理器 9-5
标准齿轮 2-6
标准公比 6-3
标准公差 1-8
标准公差等级 1-8

标准公差等级 10-1
标准化 10-1
标准夹具 6-4
标准量具 10-2
标准压力角 2-6
表 9-7
表达方案 1-9
表面粗糙度 1-8
表面粗糙度比较样块 10-4
表面粗糙度测量仪器 10-4
表面粗糙度的测量 10-4
表面建模 11-5
表面精度 12-4
表面冷作硬化 6-5
表面连接关系 1-4
表面热处理 14-6
表面质量 6-5
丙烯酸树脂 12-8
并行工程 6-9
并行口 7-3
玻璃 12-7
剥落 3-1
伯德图 17-5
薄壁特征 9-8
薄壁圆环 4-5
薄片分层叠加成形 12-8
薄片砂轮 6-2
补偿环 10-7
补偿器 17-6
补码 7-1
捕捉栅格 9-4
不导电液体 12-2
不等节距圆柱螺旋弹簧 3-5
不可分解的 17-4
不确定性 17-1
不完全退火 14-6
不稳定平衡 4-9
布局 6-8

布氏硬度 14-1
布氏硬度试验 14-1
布氏硬度值 14-1

C

擦图片 1-1
材料 14-1
材料力学 4-1
裁减窗口 11-4
裁减算法 11-4
裁剪操作 11-4
彩色图形显示适配器 11-2
菜单栏 9-1
残余奥氏体 14-6
残余应力 6-5
仓储装置 6-7
操纵离合器 3-5
操作符 7-5
槽钢 4-5
槽轮机构 2-7
槽销 3-2
草图捕捉 9-6
测力计 18-4
测量 10-2
测量对象 10-2
测量方法 10-2
测量精度 18-6
测量精确度 10-2
测量孔径 18-6
测量力 10-2
测试 18-2
测微表 10-2
叉架类零件 1-8
插齿 6-1
插齿刀/ 6-2
插齿机 15-5
插床 15-5
插床夹具 6-4
插入块 9-2
插销 6-2
插值 11-5
差动轮系 2-6
差分方程 17-2
差配继电器 18-4
拆卸画法 1-9

拆卸零件 1-9
产品模型数据交换标准 11-4
产品数据管理 11-8
铲削 6-1
长调用 7-4
长度系数 4-9
长跳转 7-4
常对数 17-5
常力的功 5-13
常矢量 5-11
常数 7-5
常数表 7-5
常微分方程 17-2
常系数系统 17-1
常用螺纹的种类和标注 1-7
常用配合 10-1
常用制图工具和仪器 1-1
超调 17-3
超高精度机床 6-3
超高速切削 6-9
超精加工 6-6
超精密 12-1
超精密加工 6-9
超静定梁 4-4
超前补偿 17-6
超声变幅杆 12-7
超声波的,超音速的 12-7
超声波加工 6-1
超声清洗 12-7
超塑性 14-5
超塑性合金 14-11
超越离合器 3-5
车床 6-3
车床夹具 6-4
车刀 5-4
车削 6-1
成分/组元 14-1
成核(形核) 14-3
成套量具 10-2
成形车刀 5-2
成形法 6-1
成形加工 6-1
成型电极 12-2
成型工序 8-1

成组技术 6-6
成组夹具 6-4
乘法指令 7-4
程序存储器 7-3
程序计数器 7-3
程序块 18-4
程序区 7-3
程序员层次交互图形系统 11-4
程序状态字 7-3
尺寸 10-1
尺寸标注 1-1
尺寸传递 10-2
尺寸公差 1-8
尺寸精度 10-7
尺寸链 6-6
尺寸稳定性 14-7
尺寸系列(滚动轴承) 3-5
齿顶圆直径 2-6
齿厚 2-6
齿厚偏差 10-11
齿厚游标卡尺 10-2
齿距累计误差 10-11
齿距偏差 10-11
齿廓啮合基本定律 2-6
齿轮 10-11
齿轮传动 3-3
齿轮传动比 5-6
齿轮传动装置 16
齿轮刀具 6-2
齿轮范成原理 16
齿轮滚刀 6-2
齿轮加工 6-1
齿轮加工机床 6-3
齿轮径向跳动检查仪 10-2
齿轮连杆机构 2-7
齿轮双面啮合检查仪 10-2
齿轮系 2-6
齿轮形插刀 16
齿轮轴 16
齿圈径向跳动 10-11
齿式联轴器 3-5
齿数比 16
齿条形插刀 16

齿向误差 10-11
齿形偏差 10-11
充电盘 18-4
充型能力 15-2
充要条件 5-4
冲裁 15-3
冲裁模 8-2
冲孔 8-2
冲量 5-11
冲压 8-1
冲压工艺 8-1
冲压工艺参数 8-1
冲压过程 8-1
冲压件 15-3
抽壳特征 9-6
出射角 17-4
出栈 7-4
初拉力 3-3
初始图形交换规范 11-4
初始应力 4-2
初值 17-3
除(去)镀(层)工艺 12-4
除指令 7-4
处于平衡状态 17-3
触发 7-6
触发器 7-6
穿透 12-6
穿越频率 17-5
传递函数 17-1
传递函数矩阵 17-2
传动比 6-3
传动角 2-4
传动精度 16
传动链 3-3
传动螺纹 10-9
传动特性 2-4
传动误差 16
传动装置 16
传感器 18-2
传输延迟 17-3
传送 7-4
传统陶瓷,普通陶瓷 14-10
串 11-3
串联 17-6
串行口 7-3

索引

窗变换 11-4
窗函数 18-3
窗口 11-4
窗视变换 11-4
床身 6-3
床头箱 6-3
创成式系统 11-7
垂直度 10-3
垂直关系 1-2
纯弯曲 4-5
磁电的 18-1
磁粉离合器 3-5
磁力测厚仪 10-2
磁力轴承 6-3
磁铁 12-6
磁性磨料电解研磨加工 12-9
磁性磨料研磨加工 12-9
磁元件 18-4
从动尺寸 9-8
从动齿轮 3-3
从动带轮 3-3
从动构件 2-2
从动环境 9-7
从动件滚子 2-5
粗大误差 10-2
粗基准 6-6
粗晶粒 14-2
粗牙普通螺纹 1-7
脆[性断]裂 14-1
脆性 14-1
脆性材料 12-1
脆性零件 12-1
脆性破坏 14-1
脆性转变温度 14-1
淬火 14-6
淬火钢 12-7
存储器 7-6
存储器扩展 7-6
存取地址 18-6
错切 11-4
错误信息 7-5

D

打断 9-3
打印预览 9-1
大齿轮 6-3
大径 10-9
大量程百分表 10-2
大柔度杆 4-9
带表卡尺 10-2
带长 3-3
带传动 3-3
带借位位减法 7-4
带进位位加指令 7-4
带进位位循环右移 7-4
带进位位循环左移 7-4
带宽 17-6
带式制动器 3-5
带状切屑 6-2
单级行星轮系 2-6
单级圆柱齿轮减速器 16
单列轴承 3-4
单排滚子链 3-3
单片机 7-2
单色的 12-5
单色相干光 12-5
单万向联轴节 2-7
单位阶跃输入 17-3
单相组织 14-4
单向离合器 3-5
单向推力轴承 3-4
单向应力状态 4-7
单项测量 10-2
单芯电缆 18-4
单一要素 10-3
单圆弧齿轮 3-3
单值量具 10-2
当量齿轮 2-6
当量摩擦角 2-3
当量摩擦系数 2-3
当量载荷 3-5
当且仅当 17-3
刀柄 6-2
刀架 6-3
刀尖 6-2
刀尖角 6-2
刀具 6-2
刀具几何形状 6-2
刀具磨损 12-1
刀具破损 6-2
刀具寿命 6-2

刀具总切削力 6-2
刀库 6-3
刀面磨损 6-2
刀片 6-2
刀体 6-2
导槽滑轮机构 5-8
导程 1-7
导电的 12-4
导轨 6-3
导套 8-2
导线 18-5
导向 8-2
导向板 8-2
导向件 6-4
导向平键 3-2
导向平键 10-10
导柱 8-2
倒角 9-3
倒棱 15-3
倒圆角 9-3
等刚度弹簧 3-5
等加速等减速运动规律 2-5
等角速度 2-2
等离子弧焊 15-4
等离子流 12-9
等离子体 12-9
等离子体加工 12-9
等离子体加速器 18-6
等强度梁 4-5
等速万向联轴节 2-7
等速运动规律 2-5
等温淬火 14-6
等温退火 14-6
等效电路 18-1
等效构件 2-3
等效力 2-3
等效力矩 2-3
等效运动方程式 2-3
等效质量 2-3
等效转动惯量 2-3
低半字节数据交换 7-4
低副 2-1
低阶系统 17-1
低碳钢 4-2
低通 18-3

笛卡尔坐标 11-4
底板 1-4
底座 6-3
地脚螺栓 3-2
地坑造型 15-2
地址总线 7-3
典型 14-12
典型零件的视图选择 1-8
点 9-2
点的三面投影 1-2
点的速度 5-5
点蚀 3-1
点阵打印机 11-2
电触式比较仪 10-2
电磁离合器 3-5
电磁吸盘 6-4
电动夹具 6-4
电动式量仪 10-2
电感 18-2
电感式测微仪 10-2
电荷 18-3
电弧焊 15-4
电化学车削 12-4
电化学珩磨 12-4
电化学加工 12-4
电化学磨削 12-4
电化学去毛刺加工 12-4
电火花加工 6-1
电火花线切割 12-3
电极 18-2
电加工工艺 12-1
电解 12-4
电解的 12-4
电解加工 6-1
电解液,电解质 12-4
电流 12-2
电流 18-3
电路 18-5
电能 12-1
电平触发 7-6
电容 18-2
电容测微仪 10-2
电容器 18-3
电压 18-3
电源 18-2
电渣焊 15-4

电子(束)流 12-6
电子束 12-6
电阻 18-2
电阻对焊 15-4
电阻焊 15-4
垫圈 3-2
吊环螺钉 3-2
掉电 7-3
叠加 1-4
碟形弹簧 3-5
碟形砂轮 6-2
丁字尺 1-1
顶点 1-3
顶角 18-3
订货到交货的周期 11-7
定点运算 18-6
定理 18-3
定量分析 17-6
定时器 7-3
定位尺寸 1-1
定位点 9-6
定位公差 10-3
定位件 6-4
定位误差 6-5
定向 9-7
定向公差 10-3
定向键 6-4
定形尺寸 1-1
定性分析 17-6
定义空间伪指令 7-5
定义字节伪指令 7-5
定义字伪指令 7-5
定制薄膜电路 12-6
定轴轮系 2-6
定轴转动 5-6
定轴转动刚体 5-12
动点 5-7
动滑动摩擦力的功 5-13
动力参数 6-3
动力学 5-1
动力学普遍定理 5-11
动力学普遍定律 5-13
动量 5-11
动量定理 5-11
动能 5-13
动能定理 5-11

动平衡 2-3
动态补偿 17-4
动态分析器 18-3
动态间隙 9-8
动态特性 17-1
动压轴承 6-3
动载荷 3-5
动坐标系 5-7
读 7-3
读图的基本要领 1-4
读组合体的视图 1-4
独立变量 17-2
独立量具 10-2
独立要求 10-3
度 11-3
端从动件 2-5
短跳转 7-4
断裂/断口 14-1
断裂边界 1-6
断裂拉伸应力 8-4
断面图 1-6
锻件图 15-3
锻压 15-3
锻造比 15-3
锻造流线 15-3
堆垛机 6-7
堆焊 15-4
堆栈 7-3
队列 11-3
对称 18-5
对称度 10-3
对称线 1-3
对称循环 3-1
对刀件 6-4
对刀块 6-4
对齐 9-4
对齐标注 9-5
对象捕捉 9-4
对象选择 9-1
对心曲柄滑块机构 2-4
镦粗 15-3
钝角 18-3
多变量系统 17-6
多次拉深 8-4
多次弯曲 8-3
多段线 9-2

多段线编辑 9-3
多功能仪器 10-2
多级行星轮系 2-6
多排滚子链 3-3
多线 9-2
多楔带 3-3
多楔滑动轴承 6-3
多芯电缆 18-4
多行文本 9-2
多油楔轴承- 6-3
多值量具 10-2
惰轮 2-6

E

额定寿命 3-5
二叉树 11-3
二次开发 9-1
二次延迟 17-3
二地址 18-4
二极管 18-1
二阶系统 17-3
二阶滞后 17-5
二进制 7-1
二进制处理 18-4
二矩式平衡方程 5-4
二力杆 4-1
二力构件 4-1
二力平衡公理 5-2
二力体 5-2
二维图形 9-1
二向应力状态 4-7
二氧化碳气体保护焊 15-4
二元合金 14-4
二元相图 14-4

F

发布绘图 9-7
发散 12-5
法向加速度 2-2
法向加速度 5-5
反馈 18-2
反码 7-1
反射板 18-4
反向电阻 18-6
反行程 2-3

反余弦 18-4
返程 18-6
返回 7-4
范成法 16
方箱法 1-5
仿形法 16
仿形机床 6-3
仿形加工 6-1
仿真 11-6
访问 7-3
放大器 18-2
放电 12-2
放影仪 18-6
放油塞 16
非 7-4
非接触测量 10-2
非晶态合金 14-11
非齐次的 17-2
非线性的 17-1
非自由体 5-2
分贝 17-5
分布参数系统 17-2
分布荷载 4-4
分层 12-8
分度头 6-4
分度头 8-6
分度圆 1-7
分段集中参数系统 17-2
分割线 9-8
分类 11-7
分离点 17-4
分离工程图 9-6
分离工序 8-1
分析函数 11-6
分型面 15-2
酚醛树脂 14-10
粉末 12-8
封闭环 10-7
峰值 18-5
峰值时间 17-3
浮点 18-6
浮动式凹模 8-3
符号 7-5
符号标志 7-3
幅角 17-4
幅值条件 17-4

索 引

幅值裕度 17-5
俯视图 1-2
辅助进位标志 7-3
辅助平面法 1-3
辅助设备 18-6
辅助时间 6-8
辅助图 1-6
辅助线 1-3
腐蚀 12-2
腐蚀的 12-2
父节点 11-3
负载能力 7-3
附加点 9-6
复变函数 17-4
复变量 17-4
复合材料 14-10
复合铰链 2-1
复合剖视图 1-6
复合样件 11-7
复频域 17-3
复位 7-3
复位状态 7-6
复杂几何形状 12-1
复杂应力状态 4-7
复制 9-3
副后刀面 6-2
副偏角 6-2
副切削刃 6-2
赋值伪指令 7-5

G

概率 6-5
概率法 10-7
概率统计的 17-2
干扰 17-3
干涉 11-5
干涉法 10-4
干涉检查 9-8
干涉仪 10-2
感应加热淬火 14-6
刚度 4-1
刚化原理 5-2
刚体 4-1
刚体定轴转动微分方程 5-12
刚体平动 5-6

刚体平面运动微分方程 5-12
刚体绕平行轴转动的合成 5-9
刚体系统 5-11
刚体系统的动量 5-11
刚体轴承动反力 5-14
刚性编码 11-7
刚性冲击 2-5
刚性联轴器 3-5
刚玉 6-2
钢尺 10-2
钢卷尺 10-2
杠杆齿轮测微仪 10-2
杠杆定律(相图) 14-4
杠杆式测微仪 10-2
高层货架 6-7
高度游标卡尺 10-2
高分子材料 14-10
高副 2-1
高阶 17-4
高精度机床 6-3
高径比(弹簧) 3-5
高速钢 6-2
哥氏加速度 2-2
各向同性 4-1
各向异性 14-2
给定精度 17-6
根 11-3
根轨迹 17-4
梗概 17-2
工步 6-6
工程数据库管理系统 11-3
工程图 9-6
工程图学 1-1
工件 6-2
工件表面 6-2
工具钢 14-7
工具栏 9-1
工位 6-6
工序 6-6
工序划分 8-1
工序余量 6-6
工艺 6-5
工艺规程 6-6

工艺基准 1-8
工艺结构 1-8
工艺卡 11-7
工艺卡片 6-6
工艺孔 6-6
工艺路线 6-9
工艺凸台 6-6
工艺装备 6-3
工作齿廓 2-6
工作高度(弹簧) 3-5
工作寄存器 7-3
工作量规 10-5
工作平面 6-2
工作台 6-4
工作位置原则 1-8
工作行程 6-6
工作循环 6-8
工作载荷 3-5
工作站 11-3
工作中心 6-9
公差带 10-1
公差带代号 10-1
公差等级 10-1
公差原则 10-3
公称直径 1-7
公法线长度 2-6
公法线长度变动 10-11
公法线平均长度偏差 10-11
公法线千分尺 10-2
公理 5-2
公制 9-1
功率 18-5
功率方程 5-13
共轭对 17-3
共轭复数 17-4
共晶反应 14-4
共析反应 14-4
钩头楔键 1-7
构件 2-1
构件的承载能力 4-9
构造几何体 9-8
构造立体几何法 11-5
固定端 4-4
固定铰支座 4-4
固化 12-8

固化 14-3
固溶体 14-2
固体润滑 3-1
故障 3-1
刮板造型 15-2
关联要素 10-3
观察孔盖 16
管理信息子系统 6-9
贯穿 1-4
惯性积 4-10
惯性矩 4-10
惯性力 2-3
惯性力矩 2-3
惯性力系 5-14
光笔 11-2
光的波长 12-8
光电 18-1
光电测扭仪 10-2
光电池 18-1
光电传感头 18-4
光电二极管 18-4
光电式检测装置 10-2
光电式量仪 10-2
光电效应 18-4
光杠 6-3
光滑极限量规 10-5
光滑接触面约束 5-2
光滑条件 4-6
光滑圆柱铰链约束 5-2
光化学加工 12-9
光敏的 12-6
光敏树脂 12-8
光敏树脂液相固化成形 12-8
光切法 10-4
光学测距仪 10-2
光学机械式量仪 10-2
光学扭簧测微计 10-2
光栅 10-2
光整加工 6-6
光子 12-5
规定画法 1-6
规格尺寸 1-9
规律 18-3
规律的 18-3
硅 12-5

辊锻 15-3
滚齿 6-1
滚齿机 15-5
滚动导轨 6-3
滚动体 3-5
滚动轴承 3-4
滚动轴承的代号 1-7
滚动轴承钢 14-7
滚筒式绘图机 11-2
滚针轴承 3-5
滚柱离合器 3-5
滚子从动件 2-5
滚子链 3-3
滚子轴承 6-3
国家标准 1-1
过渡配合 1-8
过冷 14-3
过冷奥氏体 14-6
过冷度 14-3
过盈 10-1
过盈配合 1-8
过阻尼 17-3

H

函数误差 10-2
函数约束 11-6
焊缝 1-10
焊缝符号 1-10
焊缝接头 1-10
焊后热处理 15-4
焊件 9-8
焊件剖面显示 9-7
焊接 12-6
焊接变形 15-4
焊接电弧 15-4
焊接工件 15-4
焊接接头 15-4
焊接结构 1-10
焊接图 1-10
焊接性 15-4
焊接性试验 15-4
焊接应力 15-4
焊条 15-4
焊芯 15-4
合成纤维 14-10
合成橡胶 14-10

合成粘胶剂 14-10
合金 14-2
合金钢 6-2
合金工具钢 14-7
合金元素 14-7
合金铸铁 14-8
合理标注尺寸的原则 1-8
合理的冲量 5-11
合力的功 5-13
合力矩定理 5-3
合力投影定理 5-3
荷载集度 4-4
赫尔维茨判据 17-3
黑碳化硅 6-2
黑心可锻铸铁 14-8
桁架 5-4
珩磨 6-5
横浇道 15-2
横梁 6-3
横向 9-7
轰击 12-6
红宝石 12-7
宏操作 7-5
宏定义 7-5
宏汇编程序 18-6
后刀面 6-2
后刀面磨损 6-2
后角 6-2
后视图 1-6
弧坐标 5-5
胡克定律 4-2
虎钳 6-4
虎钳 8-6
互换性 1-8
互相关 18-3
互异根 17-3
花键 3-2
花键综合量规 10-5
华氏温度 12-2
滑槽 5-2
滑动率 3-3
滑动轴承 3-4
滑键 3-2
滑块的导路 2-4
滑块联轴器 3-5
滑移齿轮 6-3

滑移方向 14-5
滑移线 14-5
化学成分 14-7
化学腐蚀 12-9
化学加工 12-9
化学抛光 12-9
化学热处理 14-6
化学热处理 15-1
化学溶液 12-4
化学铣削 12-9
环 10-7
环规 10-5
环面蜗杆 3-3
环形弹簧 3-5
环氧树脂 14-10
缓冲器 7-3
换面法 1-2
灰口铸铁 14-8
回程误差 18-1
回弹现象 8-3
回弹值 8-3
回复 14-5
回火 14-6
回火马氏体 14-6
回火索氏体 14-6
回火托氏体 14-6
回转半径 5-12
回转体 1-3
汇编程序 7-5
汇编程序文件 7-5
汇编错误码 7-5
汇编符号集 7-5
汇编结束伪指令 7-5
汇编控制 7-5
汇编命令 7-5
绘图步骤 1-1
绘图界限 9-1
绘图树 9-7
绘图仪器 1-1
混叠 18-3
混合轮系 2-6
混合模式 11-5
混杂的 12-1
锪平沉孔 1-8
活动铰支座 5-2
火花 12-2

或指令 7-4

J

机床 6-3
机床夹具 6-4
机电一体化 11-8
机动时间 6-8
机构 2-1
机构运动简图 2-1
机架 2-1
机器 2-1
机器的动能方程式 2-3
机器码传送 7-4
机器人 6-7
机器人学 11-8
机器造型 15-2
机械 2-1
机械动力装置 16
机械加工工艺结构 1-8
机械能守恒定律 5-13
机械式量仪 10-2
机械手 6-7
机械系统 16
机械效率 16
机械学 2-1
机械原理 2-1
机械制图 1-1
机械制造工艺 6-5
机械制造装备 6-3
机械装置 16
机用铰刀 6-2
机组 2-3
积分 18-5
积分常数 4-6
积分器 17-6
积聚性 1-2
积屑瘤 6-2
基本尺寸 9-5
基本额定寿命 3-5
基本功能 9-1
基本假设 4-1
基本偏差 1-8
基本视图 1-6
基本体的投影 1-3
基点法 5-8
基节偏差 10-11

索引

基孔制 1-8
基孔制配合 10-1
基面 6-2
基态 12-5
基体 9-6
基线标注 9-5
基圆 2-5
基圆直径 2-6
基轴制 1-8
基轴制配合 10-1
基准尺寸 9-6
基准孔 10-1
基准宽度 3-3
基准面 1-4
基准线 1-4
基准直径 3-3
基准制 10-1
基准轴 9-6
激发态 12-5
激光测长仪 10-2
激光打印机 11-2
激光焊 15-4
激光加工 6-1
激光扫平仪 18-4
激光束加工 12-5
激励 17-3
激振器 18-5
级数 18-5
极点 17-4
极惯性矩 4-10
极限尺寸 9-5
极限高度 3-5
极限量具 10-2
极限啮合点 2-6
极限偏差 9-5
极限强度 4-1
极限曲线 4-7
极限应力圆 4-7
极限应力圆的包络线 4-7
极限与配合 1-8
极限载荷 3-5
极值应力 4-7
极轴追踪 9-4
极坐标 5-5
极坐标图 17-5
极坐标形式 17-4

急回特性机构 2-4
急回运动 2-4
棘轮机构 2-7
集成电路 7-1
集成芯片 7-3
集体特征 9-6
集中/分散制优路 18-6
集中载荷 4-2
几何变换 11-4
几何偏心 10-11
几何图形 1-1
几何误差 6-5
几何形体 1-4
几何元素 1-3
几何作图 1-1
挤裂切屑 6-2
挤压 15-3
挤压面积 4-2
挤压破坏 4-2
计量 18-5
计量单位 10-2
计量器具 10-2
计量仪器 10-2
计量装置 10-2
计数器 7-3
计算机 7-1
计算机辅助工程分析 11-6
计算机辅助工艺规程设计 6-9
计算机辅助夹具设计 11-8
计算机辅助检测 11-8
计算机辅助设计 1-1
计算机辅助制造 11-1
计算机辅助质量控制 11-8
计算机绘图 1-1
计算机集成制造 11-8
计算机集成制造系统 6-9
计算机图形接口编码 11-4
计算机图形元文件编码 11-4
记忆功能 17-6
技术和办公协议 11-8

技术信息子系统 6-9
寄存器 7-3
加法指令 7-4
加工方法 6-1
加工精度 6-5
加工示意图 6-8
加工位置原则 1-8
加工误差 6-5
加工硬化 14-5
加工余量 6-6
加工中心 6-3
加减平衡力公理 5-2
加力(加负荷) 18-6
加热 14-6
加热喷头 12-8
加速度 5-5
加速度多边形 2-2
加速度合成定理 5-7
加速度极点 2-2
加速度计 18-6
加速度图 2-2
加速极 11-2
加速运动 5-5
加样平均值 18-6
加指令 7-4
夹紧 6-4
夹紧件 6-4
夹紧力 6-4
夹具 6-4
夹壳联轴器 3-5
夹盘 6-4
夹头 6-4
假想画法 1-9
间隙 10-1
间隙配合 1-8
间歇机构 2-7
间歇润 3-1
监视器 11-2
减光器(减声器) 18-4
减环 10-7
减速比 16
减速器 16
减速器箱座 16
减速曲线运动 5-5
减速装置 16
减振器 18-5

减指令 7-4
剪力 4-2
剪力图 4-4
剪切 4-2
剪切角 6-2
剪切面 4-2
剪切破坏 4-2
剪应变 4-2
剪应力 4-2
剪应力互等定理 4-3
检波 18-1
检波器 18-1
检测装置 10-2
检索式系统 11-7
检验夹具 10-2
简化画法 1-6
简化中心 5-4
简写符号 16
简支梁 4-4
建造模型 9-6
渐近线 17-4
渐开线齿轮 2-6
渐开线函数 2-6
渐开线花键 3-2
渐开线检查仪 10-2
渐开线蜗杆 3-3
渐开线圆柱齿轮 3-3
键 3-2
键槽 3-2
键盘 7-6
交变应力 14-1
交叉曲线 9-6
交叉线 1-2
交错轴齿轮传动 3-3
交点 1-2
交换齿轮 6-3
交换工作台 6-7
交流 18-3
交钥匙系统 11-3
浇注 15-2
浇铸系统 15-2
角度标注 9-5
角度块 10-2
角加速度 2-2
角接触推力轴承 3-4
角接触轴承 3-4

机械工程

角频率 17-5
角速度 5-6
铰削 6-1
铰刀 6-2
铰链四杆机构 2-4
阶梯剖视图 1-6
接触测量 10-2
接触面与配合面的合理结构 1-9
接触疲劳 3-1
接触线误差 10-11
接口 7-6
接口电路 7-6
节点法 5-4
节宽 3-3
节圆 10-11
结构布置 16
结构分析 2-1
结构钢,构件用钢 14-7
结合剂 6-2
结晶 14-3
截交线概念和性质 1-3
截面 12-8
截面法 4-2
截面核心 4-8
截面积 14-1
截面收缩率 4-2
解除爆炸 9-6
解理 14-5
解理断裂 14-5
解析法 5-3
解析模型 17-2
金刚石 6-2
金相组织 6-5
金属/塑料复合材料 14-10
金属化合物 14-2
金属切除率 6-2
金属切削机床 15-5
金属塑性加工 15-3
金属陶瓷 14-10
金属型铸造 15-2
筋 9-6
紧边拉力 3-3
紧定螺钉 3-2
紧固螺纹 10-9

紧密螺纹 10-9
近似展开 1-10
进给量 6-2
进给箱 6-3
进给运动 15-5
进位标志 7-3
进位位复位跳转 7-4
进位位置位跳转 7-4
经验模型 17-2
晶胞 14-2
晶格 14-2
晶核 14-3
晶界 14-2
晶界腐蚀 14-7
晶金刚石 12-5
晶粒 14-2
晶粒长大 14-3
晶粒度 14-2
晶面 14-2
晶体结构 14-2
晶体缺陷 14-2
晶体位向,晶体取向 14-2
晶向 14-2
精度 17-6
精基准 6-6
精良生产 11-8
精密锻造 15-3
精密机床 6-3
精密铸造 15-2
精确度 18-1
颈缩 4-2
径向滑动轴承 3-4
径向载荷 10-6
径向载荷系数 3-5
径向一止推滑动轴承 3-4
径向综合误差 10-11
静不定问题 4-2
静定梁 4-4
静定问题 4-2
静力关系 4-5
静力学 5-1
静平衡 2-3
静态(台架、捕获、截获)实验 18-2
静态测量 10-2
静态误差 18-5

静压轴承 6-3
静载荷 3-5
静坐标系 5-7
镜像 9-3
局部放大图 1-6
局部剖视图 1-6
局部实际尺寸 10-1
局部视图 1-6
局部自由度 2-1
矩形 9-2
矩形导轨 6-3
矩形函数 18-3
矩形花键 3-2
矩阵 11-4
句法 7-5
锯齿形螺纹 1-7
聚焦透镜 12-5
聚焦系统 11-2
决策表 11-7
决策树 11-7
绝对测量 10-2
绝对调用 7-4
绝对加速度 5-7
绝对角速度 5-9
绝对均值 18-5
绝对瞬心 2-2
绝对速度 2-2
绝对跳转 7-4
绝对温度 14-5
绝对运动 5-7
绝对坐标 9-7
绝缘材料 18-6
均匀性 4-1
均值 18-5

K

卡规 10-5
开环系统 17-1
开式链 2-1
抗尺器 18-6
抗拉强度 8-4
抗扭惯性矩 6-3
抗体 18-6
抗弯惯性矩 6-3
抗弯截面模量 4-5
可编程接口 7-6

可调夹具 6-4
可调节性 2-3
可动铰支座 4-4
可锻性 15-3
可锻铸铁 14-8
可靠性 3-1
可逆要求 10-3
可展曲面 1-10
可转为刀片 6-2
可转位车刀 6-2
课程设计 16
空操作 7-4
空格式 9-7
空间连杆机构 2-4
空间应力状态 4-7
空行程 6-6
孔 10-1
控制存取 18-6
控制电路 7-3
控制器 7-3
控制算法 17-6
控制总线 7-3
夸大画法 1-9
块 9-2
块定义 9-2
快捷菜单 9-1
快速捕捉 9-6
快速精铸 6-9
快速模具制造 6-9
快速求反工程 6-9
快速引出标注 9-5
快速原形 11-8
快速原型制造 6-9
宽带 3-3
宽度系列(滚动轴承) 3-5
辊轴支座 5-2

L

拉拔 15-3
拉床 6-3
拉刀 6-2
拉孔 6-1
拉力 4-2
拉普拉斯变换 17-2
拉伸 9-3
拉深 8-4

索引

拉深成形 8-4
拉深模 8-4
拉深系数 15-3
拉削 6-1
拉应力 4-2
拉应力区 8-4
莱氏体 14-4
劳斯稳定判据 17-3
肋 1-4
肋条 6-3
类似性 1-2
累加器 7-3
累加器高低位交换 7-4
累加器为零跳转 7-4
棱锥台 1-3
冷拔 15-3
冷变形硬化 15-3
冷冲压模具 8-1
冷处理 14-7
冷脆 14-7
冷加工 14-5
冷却 14-6
冷却曲线 14-4
冷却转变 14-6
冷作硬化 4-2
离合器 3-5
离散的 18-3
离散时间系统 17-1
离散事件 11-6
离散系统 17-2
离心调速器 2-3
离心拉力 3-3
离心铸造 15-2
离子束 12-6
理论力学 5-1
理想压杆 4-9
理想要素 10-3
理想约束反力的功 5-13
力 1-4
力的功 5-13
力的可传性 5-2
力的平行四边形法则 5-2
力对点的矩 5-3
力矩 18-5
力矩臂 18-6
力偶 5-3

力偶的功 5-13
力系 5-2
力线平移定理 5-4
力学性质 4-2
立方氮化硼 6-2
立式测长仪 10-2
立式光学比较仪 10-2
立式钻床 15-5
立体表面的展开 1-10
立体构型 1-4
立体及其表面的点 1-3
立体与立体相交 1-3
立柱 6-3
粒度 6-2
粒子 12-6
连杆曲线 2-4
连续标注 9-5
连续冷却转变曲线 14-6
连续梁 5-4
连续润 3-1
连续时间系统 17-2
连续条件 4-6
连续系统 11-6
连续性 4-1
连接链节 3-3
联立方程 17-1
联轴器 3-5
联组带 3-3
链板 3-3
链传动 3-3
链轮 3-3
链轮系 5-6
链条 3-3
链条联轴器 3-5
良导体 18-4
梁 4-4
两相组织 14-4
两箱造型 15-2
量程 18-5
量规 10-5
量化 18-3
量化误差 18-4
量角器 1-1
量具 10-2
量块 10-2
料斗 6-7

列向量 17-6
临界分切应力 14-5
临界压力 4-9
临界应力 4-9
临界应力总图 4-9
临界阻尼 17-3
临时轴 9-8
灵敏度 18-1
零、极点图 17-4
零标志 7-3
零部件测绘 1-9
零件 2-1
零点 17-4
零件测绘 1-8
零件的技术要求 1-8
零件的形状结构 1-8
零件结构的工艺性 1-8
零件图的作用和内容 1-8
零件序号 1-9
零件族 11-7
零线 10-1
溜板箱 6-3
流动性 15-2
流体摩擦 3-1
六角螺母 1-7
六角头螺栓 1-7
龙门刨床 15-5
滤波器 18-3
路径 9-8
铝 14-9
铝合金 14-9
铝青铜 14-9
绿碳化硅 6-2
轮齿 10-11
轮廓算术平均偏差 10-4
轮廓仪 10-2
轮廓中线 10-4
轮廓最大高度 10-4
轮胎式联轴器 3-5
逻辑决策 11-7
逻辑与 7-4
螺钉连接 1-7
螺距 1-7
螺栓连接 1-7
螺纹 10-9
螺纹大径 3-2

螺纹导程 3-2
螺纹的基本要素 1-7
螺纹环规 10-9
螺纹加工 6-1
螺纹紧固件 10-9
螺纹紧固件的标记 1-7
螺纹紧固件的连接形式 1-7
螺纹量规 10-9
螺纹塞规 10-9
螺纹升角 3-2
螺纹线数 1-7
螺纹小径 3-2
螺纹牙型 1-7
螺纹中径 3-2
螺旋齿轮 2-6
螺旋弹簧 3-5
螺旋副 2-1
螺旋面 1-10
螺旋线波动误差 10-11
落料 8-2

M

麻花钻 6-2
麻花钻 8-5
麻口铸铁 14-8
马氏体 14-6
埋弧焊 15-4
脉冲 7-3
脉冲响应 17-3
脉动循环 3-1
毛刺 12-4
毛坯 6-5
冒口 15-2
梅花形弹性联轴器 3-5
美国国家标准协会 11-8
美国机械工程师学会 11-8
弥散硬化 14-7
密封圈 16
密封装置 1-7
密排六方结构 14-2
面部曲线 9-6
面积(对轴)矩 4-10
面轮廓度 10-3
面向对象数据库 11-3

面心立方结构 14-2
面域 9-2
面组隐藏线移除 9-7
敏感元件 18-4
敏捷制造 11-8
名义屈服应力 4-2
明细栏 1-9
命令行 9-1
命名视图 9-4
模板 1-1
模锻 15-3
模具 8-1
模具的工作过程 8-2
模具钢 14-7
模具加工方法与刀具 8-5
模具设计 8-1
模拟的 18-2
模拟分析法 18-3
模拟记录器 18-4
模数 1-7
模数转换 7-6
模型树 9-7
模型显示 9-7
模型栅格 9-7
摩擦焊 15-4
摩擦力的功 5-13
摩擦轮传动 16
摩擦式离合器 3-5
摩擦圆 2-3
摩擦制动器 3-5
磨齿机 15-5
磨床 6-3
磨床夹具 6-4
磨具 6-2
磨粒 6-2
磨料 6-2
磨料,磨的 12-4
磨料流加工 12-9
磨料磨损 3-1
磨料悬浮液 12-7
磨损 3-1
磨损率 3-1
磨削 6-1
莫尔强度理论 4-7
目标函数 11-6
目标文件 7-5

目标文件列表 7-5

N

内槽轮机构 2-7
内棘轮机构 2-7
内浇道 15-2
内径百分表 10-2
内径千分尺 10-2
内力 4-1
内力方程 4-4
内链节 3-3
内六角圆柱头螺钉 3-2
内螺纹 3-2
内啮合 5-6
内圈 10-6
内燃机 16
内压力 4-7
内应力 6-5
内圆磨床 15-5
纳米材料 14-11
奈奎斯特图 17-5
耐磨性 3-1
难加工材料 12-1
挠度 4-6
挠曲线 4-6
挠曲线近似微分方程 4-6
能量守恒 2-3
拟定技术条件 16
拟合 11-5
逆时针 5-6
逆时针方向 17-4
逆铣 6-1
逆向工程 11-8
啮合点 2-6
啮合角 2-6
啮合线 2-6
凝固 14-3
牛顿近似法 18-6
牛头刨床 15-5
扭矩 4-3
扭矩图 4-3
扭转 4-3
扭转角 3-5

O

偶数 18-5

P

排气装置 16
排序 11-3
排样设计 8-2
派生,衍生 12-3
派生式系统 11-7
盘盖类零件 1-8
盘形凸轮 2-5
抛光 6-5
抛物线形经验公式 4-9
刨床 6-3
刨床夹具 6-4
刨削 6-1
配合 10-1
配合表面 10-1
配合尺寸 1-9
配合公差 10-1
配合公差带 10-1
配重 2-3
喷墨打印机 11-2
碰撞检查 9-8
铍青铜 14-9
皮带轮系传动 5-6
疲劳积累损伤 3-1
疲劳极限 3-1
疲劳强度 3-1
片状石墨 14-8
偏差 10-1
偏微分 18-4
偏微分方程 17-2
偏移 9-3
偏置曲柄滑块机构 2-4
漂移,偏移,偏差 17-6
频率 18-1
频率传递函数 17-5
频率响应 17-3
频率域 17-3
频谱 18-5
平板笔式绘图机 11-2
平带传动 3-3
平底从动件 2-5
平垫圈 3-2
平动刚体 5-12
平衡 2-3
平衡方程 4-2

平衡精度 2-3
平衡力系 5-2
平衡试验 2-3
平衡重量 2-3
平衡状态 5-2
平均误差 18-2
平均应力 4-2
平口虎钳 6-4
平面度 10-3
平面副 2-1
平面汇交力系 5-3
平面假设 4-2
平面力偶 5-3
平面力偶等效定理 5-3
平面力偶系 5-3
平面力系 5-3
平面立体 1-3
平面连杆机构 2-4
平面磨床 15-5
平面内的点和直线 1-2
平面平行力系 5-3
平面特殊力系 5-3
平面图形 1-1
平面弯曲 4-4
平面一般力系 5-3
平面与立体相交 1-3
平面运动方程 5-8
平面运动刚体 5-12
平稳时间(平均故障间隔时间) 18-2
平行度 10-3
平行关系 1-2
平行四边形机构 2-4
平行投影 11-4
平行投影法 1-2
平行移轴定理 4-10
平行轴齿轮传动 3-3
平行轴定理 5-12
平型砂轮 6-2
平直度测量仪 10-2
评定长度 10-4
坡口 15-4
破坏判据 4-7
剖分式滑动轴承 3-4
剖视的概念 1-6
剖视图 1-6

剖视图的类型 1-6	切割 1-4	曲面立体相交 1-3	柔性制造系统 6-8
普通带 3-3	切割法 1-5	曲线板 1-1	蠕变 14-7
普通机床 6-3	切入量 6-6	曲线平动 5-6	蠕变极限 14-7
普通平带 3-3	切舌定距 8-2	驱动尺寸 9-8	蠕墨铸铁 14-8
普通平键 1-7	切线弧 9-8	屈服极限 4-2	入口地址 7-6
普通热处理,常规热处理 14-6	切向加速度 5-5	取补 7-4	入射角 17-4
谱线展现 18-4	切向键 3-2	取样长度 10-4	入栈 7-4
	切向综合误差 10-11	全波整流 18-4	软件 7-1
Q	切削功率 6-2	全剖视图 1-6	软件包 11-3
齐次的 17-2	切削平面 6-2	全跳动 10-3	软钎焊 15-4
齐次坐标 11-4	切削热 6-2	全微分 18-4	锐角 18-3
奇偶校验 18-4	切削刃 6-2	全加速度 5-6	润滑 6-3
奇数 18-5	切削速度 6-2	全应力 4-2	润滑剂 3-1
起点偏移量 9-5	切削温度 6-2	缺省模型 9-7	润滑油 6-3
起盖螺钉 16	切削液 6-2	确定的 18-3	
起模斜度 15-2	切削用量 15-5	确定性/可靠性 18-6	**S**
起始地址 7-5	切屑 6-2		塞尺 10-2
起始地址伪指令 7-5	切屑厚度 15-5	**R**	塞规 10-5
气动夹盘 6-4	切屑宽度 15-5	热处理 14-6	三极管 18-1
气焊 15-4	切屑种类 6-2	热传导 18-4	三角板 1-1
气化 12-2	擒纵机构 2-7	热加工 14-5	三角形 9-2
气体保护焊 15-4	青铜 14-9	热能 12-6	三角形导轨 6-3
气压夹具 6-4	倾斜度 10-3	热塑性塑料 12-8	三角形函数 18-3
千分表 10-2	清零 7-4	热影响区 15-4	三矩式平衡方衡 5-4
千分尺 10-2	球化体（球状珠光体）14-6	热源 6-5	三棱锥 1-3
钎焊 15-4	球面副 2-1	人工误差 18-2	三力平衡汇交定理 5-2
牵连点 5-7	球面凸轮 2-5	人字齿轮 2-6	三视图 1-2
牵连加速度 5-7	球墨铸铁 14-8	刃具钢 14-7	三维测头 10-2
牵连角速度 5-9	球轴承 3-4	刃倾角 6-2	三维模型 12-8
牵连运动 5-7	球状石墨 14-8	刃型位错 14-2	三维实体 9-1
前刀面 6-2	区域剖面线 9-6	韧化 14-12	三向应力状态 4-7
前角 6-2	曲柄摆动导杆机构 2-4	韧性 14-1	三爪夹盘 6-3
欠阻尼 17-3	曲柄存在条件 2-4	韧性断裂 14-5	扫描 11-5
强度 4-1	曲柄导杆机构 2-4	溶解 12-4	扫描仪 11-2
强度极限 4-2	曲柄滑块机构 2-4	溶质原子 14-2	砂轮 6-2
强度理论 4-7	曲柄连杆机构 5-8	熔(化)焊 15-4	砂箱 15-2
强度设计准则 4-2	曲柄摇杆机构 2-4	熔池 15-4	砂型 15-2
强度准则 4-7	曲柄摇块机构 2-4	熔合区 15-4	砂型铸造 15-2
强化 14-7	曲柄移动导杆机构 2-4	熔模铸造 15-2	删除 9-3
强迫振动 6-5	曲柄转动导杆机构 2-4	熔丝堆积成形 12-8	闪光对焊 15-4
桥式吊车 5-7	曲杆 4-4	融化和气化 12-6	上偏差 10-1
切边 8-2	曲率 9-8	柔度 4-9	上升时间 17-3
切除量 6-6	曲率半径 4-5	柔性 6-8	上下料装置 6-7
切除特征 9-6	曲面凹模 8-3	柔性编码 11-7	烧结 12-8
切断 8-2	曲面立体 1-3	柔性冲击 2-5	舍入精度 9-5
		柔性加工单元 6-8	设备坐标系 11-4

设计变量 11-6	时钟 7-2	数据总线 7-3	速度瞬心 2-2
设计参数 11-6	实际尺寸 10-1	数控机床 6-3	速度投影法 5-8
设计基准 1-8	实际啮合线长度 2-6	数模转换 7-6	速度影像 2-2
设计周期 11-7	实际偏差 10-1	数字的 18-2	速度中心 5-9
设计自动化 18-6	实际要素 10-3	数字化仪 11-2	速率 18-5
设置参数 9-1	实例 14-12	数字控制 11-8	塑性 4-2
设置模型 9-7	实体建模 11-5	数字块 18-4	塑性变形 14-5
设置颜色 9-4	实体扫描 9-8	数字器 18-4	塑性材料 4-2
射线 9-2	实形性 1-2	数字式自动化 18-6	随机存取 18-6
深度 11-3	实验室 18-5	数字信号处理 18-5	随机误差 6-5
深度拉伸塑性变形 8-4	实验者 18-5	双变量的 18-4	随机系统 17-2
深度游标卡尺 10-2	矢量多边形 2-2	双滑块机构 2-4	随机信号储存器/只读存
深沟球轴承 10-6	矢量方程式 2-2	双列轴承 3-4	储器 18-5
深沟球轴承 3-4	矢量分析 18-3	双排滚子链 3-3	随机载荷 3-1
审阅绘图 9-7	矢量形式 5-10	双曲柄机构 2-4	随行夹具 6-4
渗氮(氮化) 14-6	使……自动化 17-1	双头螺柱 1-7	碎片 12-2
渗碳 14-6	使…偏移 12-6	双头螺柱连接 1-7	缩短 1-5
渗碳层 14-6	使用父视图造型 9-7	双万向联轴节 2-7	缩紧装置 1-7
渗碳钢 14-7	世界坐标系 11-4	双线性 18-4	所见即所得型 11-4
渗碳体 14-4	示波器 18-5	双向推力轴承 3-4	索氏体 14-6
渗碳体网状组织,网状渗	势力场 5-13	双摇杆机构 2-4	
碳体 14-4	势能 5-13	双圆弧齿轮 3-3	T
升程 2-5	视变换 11-4	双字 7-2	胎模锻 15-3
生产模式 6-9	视频带 18-6	水平仪 10-2	台式钻床 15-5
省略画法 1-6	视频图形显示适配器 11-2	水射流切割 12-9	钛 14-9
圣维南原理 4-2	视区 11-4	顺时针 5-6	钛合金 14-9
失效 3-1	手动夹具 6-4	顺铣 6-1	探试法 18-6
失效准则 4-2	手工焊 15-4	瞬时平动 5-8	碳 14-7
湿度 18-5	手用铰刀 6-2	瞬时速度 2-2	碳氮共渗 14-6
十进制 7-1	受力图 5-2	瞬时速度中心 5-8	碳当量 14-12
十进制调整 7-4	输入输出 18-3	瞬时轴 5-9	碳钢,碳素钢 14-7
十六进制 7-1	输入输出方程 17-2	瞬态 17-3	碳化物 14-7
十字滑块机构 2-4	输入文件 9-4	瞬态解 17-3	碳化物形成元素 14-7
十字滑块联轴器 3-5	输送链 3-3	瞬态模态 17-3	碳素钢 6-2
石墨 12-2	输送小车 6-7	瞬心位置 2-2	碳素工具钢 14-7
石墨化 14-8	鼠标 11-2	丝杠 6-3	弹簧 3-5
石英 12-7	术语 17-6	丝锥 6-2	弹簧垫圈 16
时变参数 17-1	数 7-1	死点 2-4	弹簧中径 3-5
时间常数 17-3	数据采集 18-2	四叉树 11-5	弹性 14-1
时间响应 17-3	数据存储器 7-3	四心近似法 1-5	弹性变形 14-1
时间域 17-3	数据地址赋值为指令 7-5	松边拉力 3-3	弹性垫圈 3-2
时间域技术 17-1	数据范围 18-6	速比 2-2	弹性极限 4-2
时效 14-6	数据缓冲器 7-3	速度 18-5	弹性夹头 6-4
时效处理 14-6	数据库管理系统 11-3	速度多边形 2-2	弹性力场 5-13
时效硬化 14-6	数据区 7-3	速度合成定理 5-7	弹性力的功 5-13
时序 7-3		速度极点 2-2	弹性流体动力润滑 3-1

索 引

弹性模量 14-1
弹性曲线 4-6
弹性套柱销联轴器 3-5
弹性应变能 4-2
镗床 15-5
镗床夹具 6-4
镗刀 6-2
镗削 6-1
掏空 9-8
陶瓷 12-5
套筒联轴器 3-5
特殊表示法 1-7
特殊功能寄存器 7-3
特殊画法 1-9
特征多边形 11-5
特征根 17-3
特征值 17-4
特种加工工艺 6-1
特种加工机床 6-3
特种加工机床 8-6
特种陶瓷 14-10
特种铸造 15-2
梯形螺纹 1-7
体积不变原则 8-4
体积应变 4-7
体心立方结构 14-2
剃齿机 15-5
替换系统标注变量 9-5
天平杆 18-6
添加剂 3-1
填充 9-2
调节 2-3
调节器 17-6
调节时间 17-3
调速器 2-3
调心滚子轴承 3-5
调心轴承 3-4
调用 7-4
调整垫片 16
调整间距 9-5
调制 18-3
调制解调器 18-3
调制器 18-3
调质 14-6
调质钢 14-7
跳动公差 10-3

跳转 7-4
铁素体 14-4
通规 10-5
通用辅助设计系统 9-1
通用机床 6-3
通用夹具 6-4
通用量仪 10-2
同步带传动 3-3
同素异构 14-3
同素异构转变 14-3
同轴度 10-3
铜 12-2
铜合金 14-9
筒形砂轮 6-2
投影基本知 1-2
投影面垂直面 1-2
投影面平行面 1-2
投影特性 1-2
投影图 1-2
透端盖 16
透明胶带 1-1
透视投影 11-4
凸凹特征 9-6
凸轮理论廓线 2-5
凸轮轮廓 2-5
凸轮轮廓上的尖点 2-5
凸缘变形区 8-4
凸缘联轴器 3-5
图板 1-1
图层 9-1
图层操作 9-4
图框 1-1
图线及其画法 1-1
图形编辑 9-1
图形单位 9-4
图形核心系统 11-4
图纸幅面 1-1
徒手绘图 1-1
团絮状石墨 14-8
推刀 6-2
推力轴承 3-4
退出 9-4
退化(变质)成 12-2
退化,变质 12-2
退火 12-6
退火处理 14-6

托氏体 14-6
椭圆 9-2

W

外部数据传送 7-4
外槽轮机构 2-7
外观标注 9-6
外棘轮机构 2-7
外径千分尺 10-2
外力 4-1
外链节 3-3
外螺纹 3-2
外啮合 5-6
外圈 10-6
外伸梁 4-4
外围设备 11-2
外形落料 8-2
外圆磨床 15-5
弯矩 4-4
弯矩图 4-4
弯曲 4-4
弯曲半径 8-3
弯曲测试 8-3
弯曲工艺 8-3
弯曲力 8-3
弯曲模 8-3
弯曲模结构 8-3
弯曲内力 4-4
弯曲应变能 4-6
弯曲中心 4-5
弯心 4-5
完全奥氏体化 14-6
完全互换法 10-7
完全退火 14-6
完全有序 14-9
完整的 12-3
碗形砂轮 6-2
万能测长仪 10-2
万能测齿仪 10-2
万向联轴节 2-7
万向联轴器 3-5
万有引力场 5-13
万有引力的功 5-13
网格 11-5
网格面 11-5
网纹滚花 1-8

网状析出 14-4
危险点 4-2
危险截面 4-2
微波电路 12-6
微处理机制优器 18-6
微分 18-5
微分方程 17-3
微分放大器 18-3
微分器 17-6
微观不平度十点高度 10-4
微机械加工 12-1
微型计算机 7-1
微震压实造型 15-2
维空间 17-6
尾座 6-3
位 7-2
位错 14-2
位地址 7-3
位地址赋值伪指令 7-5
位复位跳转 7-4
位寄存器 7-3
位图 11-2
位寻址 7-5
位移 18-5
位移边界条件 4-6
位置度 10-3
位置公差 10-3
位置量规 10-5
位置位跳转 7-4
位置位跳转并清零 7-4
温度 18-5
温度应力 4-2
文件加密 9-1
文字样式 9-4
稳定程度 17-4
稳定平衡 4-9
稳定系统 17-3
稳定性 4-1
稳弧剂 15-4
稳态精度 17-5
稳态响应 17-3
涡轮滚刀 6-2
蜗杆 2-6
蜗杆头数 3-3
蜗杆旋向 3-3

蜗轮 1-7
蜗轮滚刀 16
蜗轮节圆 3-3
蜗轮与蜗杆 1-7
卧式阿贝尔比较仪 10-2
卧式光学比较仪 10-2
无定形扫描仪号 18-4
无级变速 6-3
无限寿命设计 3-1
无心磨床 15-5
无序-有序转变 14-9
无阻尼自然频率 17-3
物理方程 4-2
物理仿真 11-6
物理关系 4-5
物料清单 6-9
物料输送系统 6-7
物料需求计划 11-8
物体系统的平衡问题 5-4
误差 18-2

X

吸收 9-6
锡青铜 14-9
铣床 6-3
铣床夹具 6-4
铣刀 6-2
铣削 6-1
系列零件设计表 9-8
系数 17-6
系数矩阵 17-6
系统误差 10-2
系统性误差 6-5
系统综合 17-6
细晶粒 14-2
细砂纸 1-1
细牙普通螺纹 1-7
下偏差 10-1
先验知识 17-1
纤维纸板垫片 16
纤维状组织 14-5
显示处理器 11-2
显示剖面线 9-7
显示器 7-6
显微组织变化,显微结构变化 14-4

现代控制理论 17-1
线段分析 1-1
线宽 9-4
线框 11-5
线框建模 11-5
线轮廓度 10-3
线面分析 1-4
线速度 5-6
线形 9-4
线形比例 9-4
线性的 18-1
线性向量空间 17-1
线应变 4-2
相 14-6
相变 14-4
相变点 14-6
相变点,转变点 14-4
相当应力 4-7
相对测量 10-2
相对加速度 5-7
相对视图 9-8
相对瞬心 2-2
相对速度 2-2
相对弯曲半径 8-3
相对稳定性 17-5
相对误差、比例误差 18-2
相对运动 5-7
相对坐标 9-7
相反形状的 12-2
相干性 12-5
相贯体 1-3
相贯线的概念和性质 1-3
相交 17-4
相交关系 1-2
相交轴齿轮传动 3-3
相角条件 17-4
相角裕度 17-5
相平面方法 17-1
相似原则 8-4
相图 14-4
相位超前 17-4
相位滞后 17-4
箱盖 16
箱体类零件 1-8
响应 7-6
向量 17-6

向视图 1-6
向心滚子轴承 3-4
向心轴承 1-7
像素 11-2
橡胶 14-10
削边销 6-4
消失模铸造 15-2
消隐 11-5
销 3-2
小变形 4-1
小齿轮 6-3
小径 10-9
小型计算机 7-1
校对量规 10-5
校核强度 4-2
楔键 3-2
斜齿大齿轮 16
斜齿轮 1-7
斜二等轴测 1-5
斜角切削 6-2
斜截面 4-2
斜率 17-6
斜剖视图 1-6
斜视图 1-6
斜投影法 1-2
斜弯曲 4-8
斜线 1-2
谐波齿轮传动 2-6
写 7-3
卸载定理 4-2
芯盒 15-2
芯轴 6-4
信号 18-1
信号发生器 18-5
信息 18-1
行程速度变化系数 2-4
行星齿轮 2-6
行星齿轮机构 5-8
行星轮系 2-6
行星转臂 2-6
形变带 14-5
形体分析 1-4
形心 4-5
形心主惯性矩 4-10
形心主轴 4-10
形状改变比能(第四强

度)理论 4-7
形状改变能密度 4-7
形状公差 10-3
形状和位置公差 1-8
形状记忆合金 14-11
形状记忆效应 14-11
形状特征原则 1-8
型腔 9-8
性能 14-1
性能指标 17-3
修改特性 9-4
修改文本 9-3
修剪 9-3
修正的,补偿的 18-4
虚拟现实 11-2
虚拟制造 6-9
虚线 17-4
虚约束 2-1
许可载荷 4-2
许可转角 4-6
悬臂梁 4-4
旋绕比(弹簧) 3-5
旋向 1-7
旋转 9-3
旋转精度 18-6
旋转剖视图 1-6
选材 14-12
选取实例 9-7
选项卡 9-1
选择性激光粉末烧结成
形 12-8
渲染 9-7
循环控制指令 7-4
循环润 3-1
循环特性 3-1
循环右移 7-4
循环载荷 3-1
循环左移 7-4

Y

压(力)焊 15-4
压板 6-3
压电 18-1
压杆的稳定容许应力 4-9
压杆的稳定条件 4-9
压杆失稳 4-9

索 引

压肩 15-3
压力机/冲床 8-1
压力角 1-7
压力铸造 15-2
压钳口 15-3
牙嵌式联轴器 3-5
牙嵌式制动器 3-5
牙型角 3-2
氩弧焊 15-4
延伸 9-3
延伸公差带 10-3
延伸率 4-2
研磨 6-6
演绎法 18-6
验收量规 10-5
燕尾形导轨 6-3
扬声器 18-4
阳极 12-2
仰视图 1-6
氧乙炔焊 15-4
样板 9-1
样本 18-5
样本均值 18-6
样条 11-5
样条曲线 9-2
摇臂钻床 15-5
摇块机构 2-4
药皮 15-4
冶金的 12-2
页面 9-7
液晶显示 11-2
液体动压滑动轴承 3-4
液体静力润滑 3-1
液体静压滑动轴承 3-4
液体润滑 3-1
液压夹具 6-4
液压夹盘 6-4
液压减振器油压缓冲装置 18-5
一般位置直面 1-2
一层一层地 12-8
一齿径向综合误差 10-11
一齿切向综合误差 10-11
一点的应力状态 4-7
一阶系统 17-3
一阶滞后 17-5

一矩式平衡方程 5-4
移出断面图 1-6
移动 9-3
移动从动件 2-5
移动副 2-1
已加工表面 6-5
已加工表面质量 6-2
异或 7-4
译码器 7-3
溢出标志 7-3
因变量 17-2
阴极 12-2
阴极脉冲调制的 18-4
阴极射线管 11-2
音频带 18-6
引出端 18-4
引脚 7-3
引力 18-6
应变能密度 4-7
应力 4-2
应力分量 4-7
应力腐蚀,晶间开裂 14-7
应力集中 4-2
应力圆 4-7
英制 9-1
盈亏功 2-3
荧光屏 11-2
影像法 2-2
硬脆材料 12-1
硬度 6-2
硬化层深度 14-6
硬件 7-1
硬钎焊 15-4
硬质合金 6-2
用户界面 17-6
用户界面管理系统 11-4
优点,指标,准则 17-6
优化 11-6
优化方法 11-6
优先级 7-6
优先配合 10-1
优先数系 10-1
由装配图拆画零件图 1-9
油标 16
油槽 3-4
油环润滑 3-1

油孔 3-4
油雾润滑 3-1
油浴润滑 3-1
游标卡尺 10-2
游标量具 10-2
游标式量仪 10-2
有机变速 6-3
有势力的功 5-13
有限寿命设计 3-1
有限元 11-6
有限元法 11-6
有效 18-5
有效拉力 3-3
有效力 2-3
有效圈数 3-5
有阻尼自然频率 17-3
右视图 1-6
右手定则 5-6
右移 7-4
余弦 18-4
余弦加速度运动规律 2-5
语言测听计 18-6
语音识别系统 11-2
语音系统 11-2
预锻 15-3
元冲量 5-11
元功 5-13
元件 18-5
原材料 6-5
原点 17-3
原动机 16
原理图 17-6
原码 7-1
原始单元体 4-7
原子排列 14-2
原子喷射 12-6
原子团 14-3
原子位错,原子位移 14-2
圆 9-2
圆带传动 3-3
圆度 10-3
圆度仪 10-2
圆弧 9-2
圆弧连接 1-1
圆弧圆柱齿轮 3-3
圆弧圆柱蜗杆 3-3

圆环 9-2
圆截面 4-5
圆螺母 3-2
圆盘凸轮机构 5-7
圆跳动 10-3
圆柱齿轮 1-7
圆柱度 10-3
圆柱副 2-1
圆柱管螺纹 1-7
圆柱滚子轴承 3-4
圆柱滚子轴承 10-6
圆柱铰链 5-2
圆柱螺纹 3-2
圆柱螺旋拉伸弹簧 3-5
圆柱螺旋扭转弹簧 3-5
圆柱螺旋压缩弹簧 3-5
圆柱凸轮 2-5
圆柱蜗杆 3-3
圆柱销 1-7
圆柱形导轨 6-3
圆锥 10-8
圆锥顶 1-3
圆锥公差 10-8
圆锥管螺纹 1-7
圆锥滚子轴承 3-4
圆锥角 10-8
圆锥量规 10-8
圆锥螺纹 3-2
圆锥配合 10-8
圆锥塞规 10-8
圆锥销 1-7
圆锥形凸轮 2-5
源文件 7-5
源文件列表 7-5
约束 5-2
约束反力 5-2
约束反力系 5-14
约束条件 17-4
约束系数 4-9
月牙洼磨损 6-2
匀变速转动 5-6
匀速曲线运动 5-5
匀速转动 5-6
运动参数 6-3
运动方程 5-5
运动副 2-1

运动副元素 2-1
运动轨迹 5-5
运动偏心 10-11
运动确定性 2-1
运动失真 2-5
运动学 5-1
运算器 7-3

Z

载波 18-3
载荷叠加 4-6
再结晶 14-5
在制品 11-7
造型 15-2
噪声去除器 18-5
噪声衰减 18-6
噪声听度计 18-6
噪音 18-5
增环 10-7
增强 14-10
增速比 16
增速装置 16
增益 17-4
增益参数 17-4
轧辊 15-3
窄带 3-3
展成法 6-1
展开图 1-10
栈 11-3
找正 6-4
折光天线 18-3
折弯标注 9-5
针描法 10-4
针状马氏体 14-6
真空腔(室) 12-6
真实应力 14-1
阵列 9-3
阵列特征 9-6
振荡 17-3
振荡器 7-3
振动 6-5
振动拾取 18-5
振幅 17-3
振荡器(交流发电机) 18-2
整流 18-1

整流器 18-1
整体式滑动轴承 3-4
正齿轮 1-7
正等轴测图 1-5
正多边形 9-2
正火 14-6
正交功能 9-4
正交坐标系 11-4
正六棱柱 1-3
正四棱锥 1-3
正态分布 6-5
正投影法 1-2
正弦机构 2-4
正弦加速度运动规律 2-5
正应力 4-2
正应力分布 4-5
支撑板 1-4
支承板 6-4
支承件 6-3
支点位移条件 4-6
支架 1-4
枝(状)晶 14-3
枝晶长大 14-3
知识获取系统 11-7
知识库 11-7
执行汇编程序 7-5
执行机构 18-2
直齿轮 10-11
直齿圆锥齿轮 2-6
直浇道 15-2
直角尺 10-2
直角切削 6-2
直角投影定理 1-2
直角坐标形式 5-10
直径标注 9-5
直径系列(滚动轴承) 3-5
直流 18-3
直流(电) 12-2
直纹滚花 1-8
直线 9-2
直线标注 9-5
直线裁剪 11-4
直线的实长和倾角 1-2
直线的投影特性 1-2
直线度 10-3
直线拉长 9-3

直线平动 5-6
止规 10-5
止推滑动轴承 3-4
指定模板 9-7
指令 7-5
指令周期 7-2
指数的 17-3
指针 11-3
制动器 3-5
制图基本规格 1-1
制芯 15-2
制造单元 11-8
制造哲理 6-9
制造资源计划 11-8
制造自动化协议 11-8
质点的达朗伯原理 5-14
质点的动量 5-11
质点的动量定理 5-11
质点的动量矩 5-12
质点的动量矩定理 5-12
质点的动量守恒 5-11
质点的动能 5-13
质点的动能定理 5-13
质点系的达朗伯原理 5-14
质点系的动量 5-11
质点系的动量定理 5-11
质点系的动量矩 5-12
质点系的动量矩定理 5-12
质点系的动量守恒 5-11
质点系的动能 5-13
质点系的动能定理 5-13
质点系的质心 5-11
质点系内力的功 5-13
质点系相对质心的动量矩定理 5-12
质径积 2-5
质量保证子系统 6-9
质心,心形曲线 17-4
质心运动定理 5-11
质心运动守恒定律 5-11
智能尺寸 9-6
智能化 9-1
滞后补偿 17-6
置位 7-3

中断 7-6
中断返回 7-4
中断请求 7-6
中断源 7-3
中间平面 3-3
中径 10-9
中小柔度杆 4-9
中心标注 9-5
中心符号线 9-8
中心架 6-3
中心距 1-4
中心孔 1-4
中心投影法 1-2
中心线 9-6
中心线弯曲角 8-3
中心钻 6-2
中性层 4-5
中性轴 4-5
中性轴方程 4-8
中央处理器 7-2
终端 11-2
终端负载 18-2
终锻 15-3
重叠 18-3
重复精度 18-6
重合断面图 1-6
重建模型 9-8
重力场 5-13
重力的功 5-13
重命名 9-4
重影 1-2
周期 18-3
周期性 18-3
周转轮系 2-6
轴 4-3
轴测剖视图 1-5
轴测投影面 1-5
轴测图 1-5
轴测图的分类 1-5
轴承 6-3
轴承衬 3-4
轴承承载能力 3-4
轴承径向载荷 3-4
轴承宽度 3-5
轴承内径 3-5
轴承外径 3-5

索引

轴承系列（滚动轴承）3-5
轴承压强 3-4
轴承轴向载荷 3-4
轴端挡圈 16
轴间角 1-5
轴肩挡圈 16
轴交角 3-3
轴力 4-2
轴力图 4-2
轴套 16
轴套类零件 1-8
轴瓦 3-4
轴向拉伸 4-2
轴向伸缩系数 1-5
轴向压缩 4-2
轴向载荷 10-6
轴向载荷系数 3-5
珠光体 14-4
主参数 6-3
主单元体 4-7
主动测量 10-2
主动齿轮 3-3
主动带轮 3-3
主动的、有源的 17-4
主动构件 2-2
主动力 5-2
主动力系 5-14
主惯性矩 4-10
主惯性轴 4-5
主后刀面 6-2
主极点 17-3
主矩 5-4
主面 4-7
主偏角 6-2
主频 7-2
主切削刃 6-2
主生产计划子系统 6-9
主矢 5-4
主视图 1-2
主应力 4-7
主应力迹线 4-7
主运动 6-1
主轴 6-3
主轴部件 6-3
主轴材料 6-3
主轴结构 6-3

主轴箱 6-3
主轴转速 6-3
属性定义 9-4
属性匹配 9-4
注释 7-5
注释绘图 9-7
柱状晶 14-3
柱坐标 5-5
铸件 15-2
铸铁 4-2
铸造 15-2
铸造工艺结构 1-8
铸造零件的工艺结构 1-8
铸造缺陷 15-2
铸造性能 15-2
铸造组织 15-2
专家系统 11-8
专门化机床 6-3
专用机床 6-3
专用夹具 6-4
专用量仪 10-2
转变 14-6
转动方程 5-6
转动副 2-1
转角 4-6
转力矩 18-5
转速图 6-3
转向线 1-3
转折频率 17-5
转轴定理 4-10
装配 16
装配尺寸 1-9
装配方法 6-6
装配工艺结构 1-9
装配关系和工作原理 1-9
装配结构的合理性 1-9
装配精度 6-6
装配示意图 1-9
装配体 9-6
装配体配合 9-6
装配误差 6-6
装配应力 4-2
装饰螺纹线 9-8
状态变量 11-6
状态变量方法 17-5
状态栏 9-1

锥齿轮 1-7
锥度 10-8
准时生产 11-8
着色 9-7
子节点 11-3
子树 11-3
子装配体 9-6
子装配体 9-8
字 7-2
字节 7-2
字节地址 7-3
字节寄存器 7-3
字节交换 7-4
字节寻址 7-5
自定义设置 9-4
自动编程工具 11-8
自动调心滑动轴承 3-4
自动过渡 9-6
自动化造型 15-2
自动化制造子系统 6-9
自动换刀装置 6-3
自动机床 6-3
自动检测系统 10-2
自动检测装置 10-2
自动取样器 18-6
自动扫描 18-6
自动生产线 6-8
自攻螺钉 3-2
自激振动 6-5
自谱 18-6
自然环境 16
自然形式 5-10
自然轴系 5-5
自润滑滑动轴承 3-4
自上而下的设计 9-8
自锁 2-3
自相关 18-3
自由度 6-4
自由度数 2-1
自由锻 15-3
自由高度（弹簧）3-5
自由体 6-4
自由振动 6-5
综合测量 10-2
总圈数 3-5
总衰减 18-6

总体尺寸 1-9
总体刚度矩阵 11-6
总体平均 18-6
纵向 9-7
阻断 17-1
阻尼比 17-3
阻尼器 18-6
组成环 10-7
组合变形 4-8
组合弹簧 3-5
组合机床 6-3
组合机床自动线 6-8
组合体 1-4
组合体的尺寸标注 1-4
组合体的分析 1-4
组合体的三视图 1-4
组合体的形成方式 1-4
组合体视图的画法 1-4
组合图形 4-10
组合相贯线 1-3
组合夹具 6-4
组织 6-2
钻床 15-5
钻床夹具 6-4
钻套 6-4
钻削 8-5
最大过盈 10-1
最大极限尺寸 10-1
最大间隙 10-1
最大剪应力（第三强度）理论 4-7
最大拉应力（第一强度）理论 4-7
最大伸长线应变（第二强度）理论 4-7
最大实体极限 10-1
最大实体要求 10-3
最大系统 7-5
最小二乘逼近 18-6
最小过盈 10-1
最小极限尺寸 10-1
最小间隙 10-1
最小实体极限 10-1
最小实体要求 10-3
最小系统 7-5
最小相位 17-5

最优控制 17-1
左视图 1-2
左移 7-4

作用力与反作用力定律 5-2

作用线 5-4
坐标标注 9-5

坐标法 1-5
坐标系 5-7